Storey's Guide to
RAISING TURKEYS

Storey's Guide to
RAISING TURKEYS

LEONARD S. MERCIA

With revisions by Jesse Grimes, Ph.D.
Department of Poultry Science
North Carolina State University
Raleigh, North Carolina

The mission of Storey Publishing is to serve our customers by publishing practical information that encourages personal independence in harmony with the environment.

Edited by Marie Salter and Deborah Burns
Copyedited by Mary Boylan
Cover design by Renelle Moser
Front cover photograph © Grant Heilman from Grant Heilman Photography, Inc.
Back cover photograph by Corbis Images
Interior photographs by Jesse Grimes
Illustrations by Elayne Sears, except those on pages 9, 10, 16 (bottom), 20, 21, 42, 44 (top), 45, 50–51, 79, 86, 87, and 89
Series design by Mark Tomasi
Text production by Erin Lincourt and Deborah Daly
Indexed by Susan Olason/Indexes & Knowledge Maps

Storey's Guide to Raising Turkeys was previously published under the title *Raising Your Own Turkeys*. This new edition has been expanded by 64 pages. All of the information in the previous edition was reviewed and revised for this new text, which offers the most comprehensive and up-to-date information available on the topic of raising turkeys.

Printed in the United States by Versa Press
10 9 8 7 6 5

Library of Congress Cataloging-in-Publication Data

Mercia, Leonard S.
Storey's guide to raising turkeys / Leonard S. Mercia.
p. cm.
Includes index.
ISBN 1-58017-261-X (alk. paper)
1. Turkeys. I. Title: Guide to raising turkeys. II. Title.
SF507.M48 2000
636.5'92—dc21 00-057342

Contents

PREFACE

In recent years, there has been considerable interest in raising small flocks of poultry. This trend actually started back in the 1970s and continues to this day. While working as Extension Poultry Specialist at the University of Vermont during the 1970s, I noticed a rapid increase in the number of office contacts asking questions about raising different types of poultry, including turkeys. There were hundreds of requests annually, so I began to analyze the types of questions asked and the information requested.

It quickly became apparent that we did not have available adequate educational material in the form of handouts to respond to all questions. The only texts available at the time were written during the pre–World War II era and were not current enough to be satisfactory. At that point, I decided to write a reference manual on general poultry, and soon thereafter *Raising Poultry the Modern Way* (1975; now *Storey's Guide to Raising Poultry)* was published. That book was followed by *Raising Your Own Turkeys*, which has been completely updated for this new edition, *Storey's Guide to Raising Turkeys*.

You might choose to raise a small flock of turkeys for any number of reasons. Wanting to raise turkeys for the family table is one possibility. A small flock of turkeys can also provide a family member with added responsibility, a hobby, or even a source of extra income. Whatever the reason, if the project is to be successful, you must have a good understanding of all aspects of rearing, processing, and marketing turkeys. This book has been written with those needs in mind.

Storey's Guide to Raising Turkeys includes information on how to get started with a turkey project so as to avoid some of the possible pitfalls. It contains information on choosing breeds and varieties, managing growing birds, housing and equipment requirements, feeding requirements, flock health, incubation, processing, and marketing. Plans are also included for making your own turkey houses and equipment.

The book concludes with helpful lists of reading materials, diagnostic laboratories, Cooperative Extension System offices, associations, sources of supplies and equipment, and a glossary, making this an invaluable reference for turkey raisers at all levels.

ONE

GETTING STARTED

The turkey: an all-American bird! The turkey was here long before the Europeans ever came to the New World. Estimates put the turkey's ancestors in North America somewhere around 1.8 to 5 million years ago. Native Americans ate turkey well before the Europeans' arrival, and not just once or twice a year but year-round. They knew that the turkey was a great food resource. As a matter of fact, the natives in Old Mexico and the southwestern United States domesticated the turkey and were the first turkey farmers. Other Native Americans hunted the bird across its range. The turkey not only was a source of food for the natives but also provided clothing, tools, and weapons; contributed to their ceremonies; and had numerous other miscellaneous uses. When the Europeans took the turkey back home, it was given such high honors that in many cases only the elite and royalty could eat it. When the colonists came to the New World, they were surprised to find the turkey already here, but were quick to add it to their food supply.

Fortunately, we do not have to worry about where our turkeys come from. Farmers in the United States produce 275 to 300 million turkeys each year, enough to supply U.S consumers and export turkey products to other countries. Turkeys are fed wholesome diets made up of corn, soybean meal, wheat, and other ingredients, such as vitamins and minerals.

Successful and caring turkey producers spend a great deal of energy, time, effort, and money to ensure that their turkeys are given the best care possible. This means that people can purchase turkey products knowing that they are providing their families with one of the most nutritious and wholesome foods available. Turkey has always been popular on our dinner table during Thanksgiving and other holidays; however, maybe we should take a hint from the first Americans and use the turkey as a wholesome food year-round.

Why Raise Turkeys?

Even though turkeys are raised in large numbers by the turkey industry, there are several reasons why people might enjoy raising small flocks themselves. Having fresh turkey on your dinner table all year is only one reason to raise a small flock of these wonderful birds. Raising turkeys can also be an excellent family or youth project and a great way for anyone to develop an

Raising turkeys can be an excellent family or youth project.

understanding of live-animal management. In addition, you might be able to build a profitable market for at least a small number of turkeys. Many consumers are becoming much more interested in buying their turkey from "organic" or "free-range" sources; others just wish to become a little closer to the land by purchasing their food directly from farmers.

Turkeys are interesting birds to raise. Members of 4-H and other youth clubs seem to enjoy turkey projects as much as, and maybe more than, other livestock projects. They have good results with them, too. Turkeys are fun to raise because they are docile with people and, given the opportunity, will follow you and observe your every move. Not only are they curious; at times they're protective as well. When strangers approach your home, barnyard birds will alert you in their noisy, gobbly fashion.

Special Considerations

Turkeys are not as difficult to raise as many people think. However, they do require special care to get them off to a good start. Sometimes they are a little slow in learning to eat and drink. Turkeys should be isolated from chickens and other poultry to prevent many diseases. It is important that turkey *poults* (that is, young turkeys) be kept warm and dry during the first few weeks after hatching; this time is called the *brooding period*. If you start with good stock and provide good feed, housing, and husbandry, you can raise turkeys successfully.

However, before launching into production, even on a small scale, be aware of the costs. Day-old turkey poults are quite expensive, and they consume a considerable amount of feed; thus, the cost of producing full-grown market turkeys is quite high.

Before starting a flock, check local laws and ordinances. Zoning regulations in some areas prohibit keeping poultry of any kind. If you live close to neighbors, keep in mind that noise, odor, and possibly fly problems are associated with raising turkeys.

Estimated Per-Bird Costs of Raising Heavy Roaster Turkeys

ITEM	COST
Day-old poult	$ 1.50– 2.00
Feed (75 lbs. [34 kg])*	10.50–12.00
Brooding, electricity, litter, miscellaneous	0.45– 0.60
Total	$12.45–14.60

Assumptions: Small flock with relatively high feed costs; poult costs are usually high; all feed purchased; no labor, housing equipment, interest, processing, or marketing costs are included. (Costs based on 2000 figures.)

*Feed consumption depends on the bird's variety, strain, and age at processing.

Varieties

Several domestic varieties have been developed from the wild turkey. The Large White turkey is the most important for the commercial turkey industry. The Medium White turkey is popular in other countries and should be considered a viable option for consumption by consumers and production by turkey growers in the United States. The Medium White reaches market age earlier than the Large White and provides a plump, whole carcass that is the right size for modern families. The Medium White may be difficult for the small-flock producer to find because it is being developed by the commercial industry with few outlets for the small-flock producer. The Beltsville Small White (sometimes called the Beltsville White) was popular at one time, but this bird has fallen out of favor and is now very difficult to find.

During the early years of development of the commercial turkey industry, the Broad-Breasted Bronze was the most popular variety. The Bronze has a good growth rate, conformation (or meatiness), feed conversion, and most of the other qualities demanded by the turkey industry; however, it has dark pinfeathers that, if left on, might detract from the dressed appearance. This disadvantage led to the gradual replacement of the Bronze with white birds by the commercial turkey industry. However, the Broad-Breasted Bronze is a heavy-type bird that is still excellent for hobby flock producers, even though it may take extra time to remove the dark pinfeathers. It may be more available to these producers and may be a hardier bird than the commercial Large White. There are several other types of turkeys available to producers who are not necessarily interested in economic traits (that is, fast growth, conformation, efficient feed consumption, white feathers, and larger body size). These varieties include White Holland, Black, Royal Palm, Bourbon Red, and Narragansett, all of which are beautiful birds and should be considered for rearing, especially by those interested in hobby flocks.

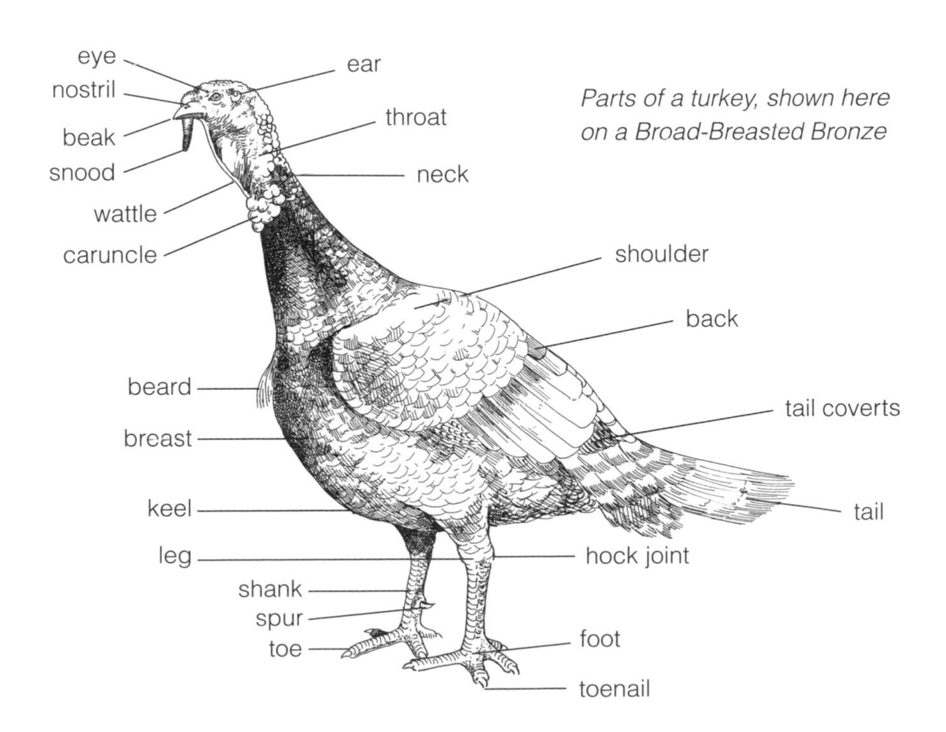

Parts of a turkey, shown here on a Broad-Breasted Bronze

Hens of the fast-growing, heavy roaster turkey species, such as the Large White and the Broad-Breasted Bronze, usually reach a live weight of about 15 pounds (6.8 kg) at 14 weeks of age; toms weigh approximately 30 pounds (13.6 kg) at 18 weeks. Your birds can be smaller or larger depending on the age at which they are slaughtered.

Important Note

It's hard to beat the flavor of a well-finished, mature turkey, especially one that is fresh. However, you may grow tired of turkey in time. Remember: Don't raise more birds than you can eat and enjoy unless the surplus can be marketed profitably. In some cases, well-finished, fresh-dressed turkeys can be sold at prices substantially above their production costs.

Buying Poults

There are several ways to get started with a few poults, but the simplest and least expensive method is to buy the poults directly from a hatchery on the day of hatching. While working with breeder turkeys is interesting and satisfying, keeping breeders and hatching the eggs is an expensive, time-consuming way to start a flock, especially for people interested in building a profitable business. Turkey poults from commercial Large White and Broad-Breasted Bronze strains provide the best performance.

New poults should originate from U.S. breeder flocks that are free of pullorum, typhoid, and preferably *Mycoplasma* organisms, which cause infectious sinusitis and synovitis. Consult a state poultry specialist, a county agricultural Extension agent, a commercial producer, or some other knowledgeable person for advice on good sources for turkey poults in your area.

Place your order for poults well in advance of the delivery date so you can be sure to get the stock you want. It is sometimes difficult to obtain small lots of turkeys delivered from the hatchery. However, it may be possible to pick up the turkeys at the hatchery or perhaps to have a nearby independent producer order a few extra poults for you. Frequently, day-old or started poults can be purchased from local feed and farm-supply outlets or from specialized dealers who handle small lots of chicks, poults, and other types of birds. Some mail-order companies that specialize in poultry and game-bird production also provide fertile eggs or day-old poults. This may be the best buy for small-flock producers.

The best time to start the small turkey flock is in late May or early June. Starting poults at that time enables you to grow them to the desired market weights just prior to the traditional holiday season, when the demand for turkey is strongest. It also avoids starting poults during the coldest season of the year, which can make brooding much more difficult.

Day-old poults can be purchased from local feed and farm-supply outlets and from some mail-order companies.

Housing

If you have a small turkey flock that is usually started in the warm months of the year, housing does not have to be fancy. However, the brooder house should be a reasonably well-constructed building that can be readily ventilated. If a small building is not available, perhaps a pen space within a larger building can be provided.

The area should have good floors that can be easily cleaned and disinfected. Concrete floors are preferred, but wood floors are acceptable.

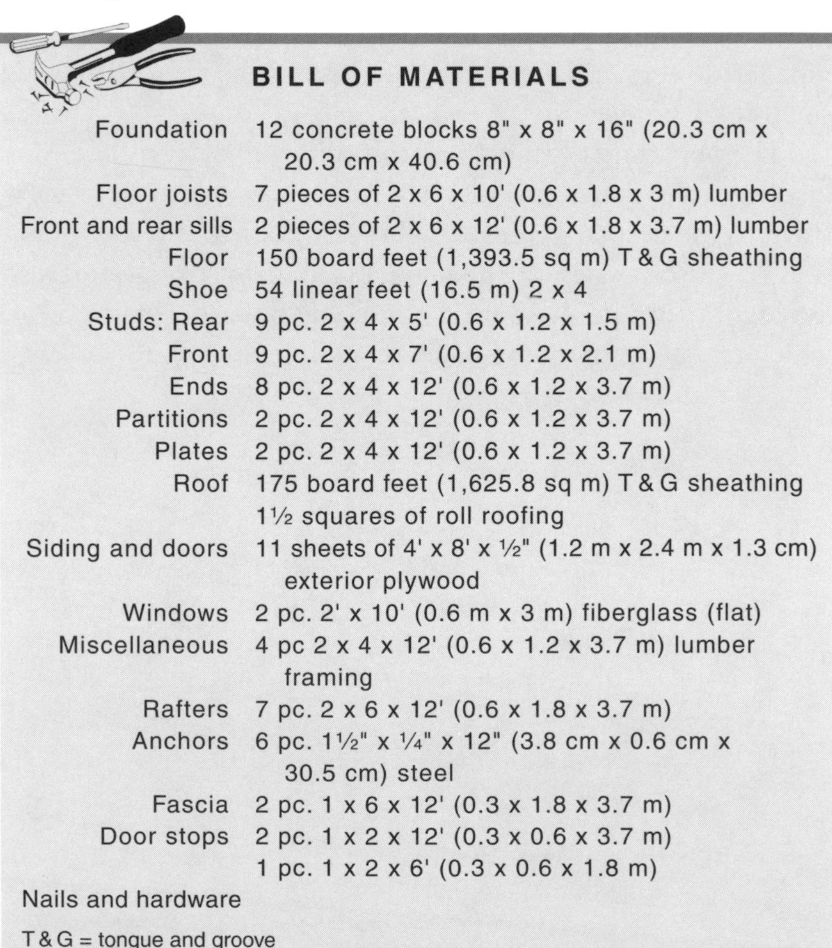

BILL OF MATERIALS

Foundation	12 concrete blocks 8" x 8" x 16" (20.3 cm x 20.3 cm x 40.6 cm)
Floor joists	7 pieces of 2 x 6 x 10' (0.6 x 1.8 x 3 m) lumber
Front and rear sills	2 pieces of 2 x 6 x 12' (0.6 x 1.8 x 3.7 m) lumber
Floor	150 board feet (1,393.5 sq m) T & G sheathing
Shoe	54 linear feet (16.5 m) 2 x 4
Studs: Rear	9 pc. 2 x 4 x 5' (0.6 x 1.2 x 1.5 m)
Front	9 pc. 2 x 4 x 7' (0.6 x 1.2 x 2.1 m)
Ends	8 pc. 2 x 4 x 12' (0.6 x 1.2 x 3.7 m)
Partitions	2 pc. 2 x 4 x 12' (0.6 x 1.2 x 3.7 m)
Plates	2 pc. 2 x 4 x 12' (0.6 x 1.2 x 3.7 m)
Roof	175 board feet (1,625.8 sq m) T & G sheathing 1½ squares of roll roofing
Siding and doors	11 sheets of 4' x 8' x ½" (1.2 m x 2.4 m x 1.3 cm) exterior plywood
Windows	2 pc. 2' x 10' (0.6 m x 3 m) fiberglass (flat)
Miscellaneous	4 pc 2 x 4 x 12' (0.6 x 1.2 x 3.7 m) lumber framing
Rafters	7 pc. 2 x 6 x 12' (0.6 x 1.8 x 3.7 m)
Anchors	6 pc. 1½" x ¼" x 12" (3.8 cm x 0.6 cm x 30.5 cm) steel
Fascia	2 pc. 1 x 6 x 12' (0.3 x 1.8 x 3.7 m)
Door stops	2 pc. 1 x 2 x 12' (0.3 x 0.6 x 3.7 m) 1 pc. 1 x 2 x 6' (0.3 x 0.6 x 1.8 m)
Nails and hardware	

T & G = tongue and groove
o.c. = on center

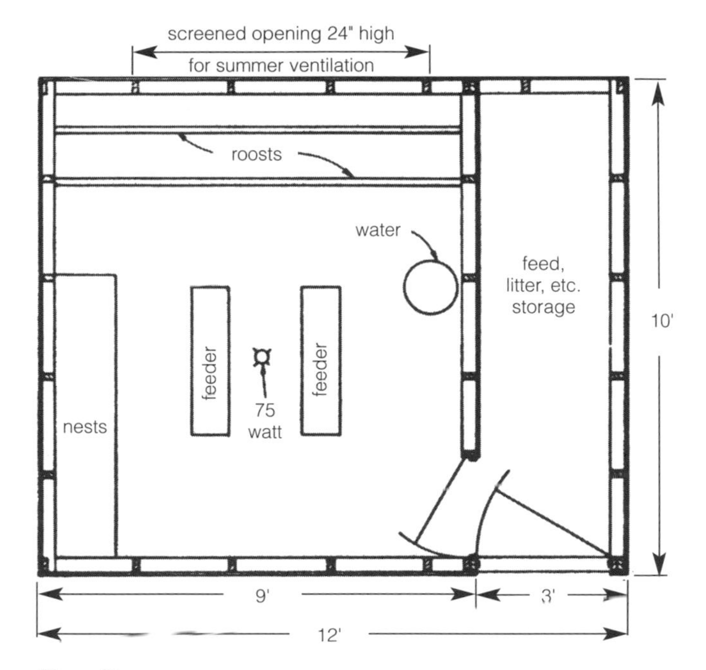

Floor Plan

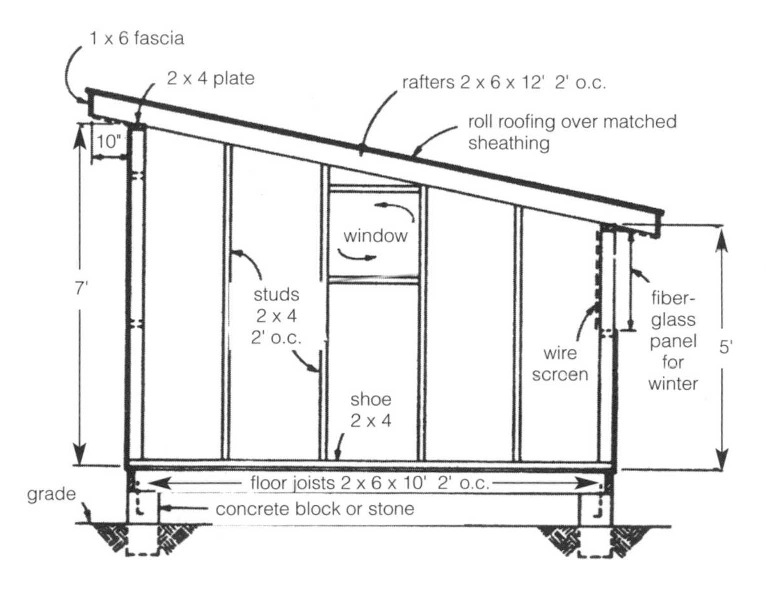

Side Elevation Framing

Plan and elevations for a 10- x 12-foot (3.1 x 3.7 m) poultry house. (Note: *Consult local health and building code authorities before starting construction.)*

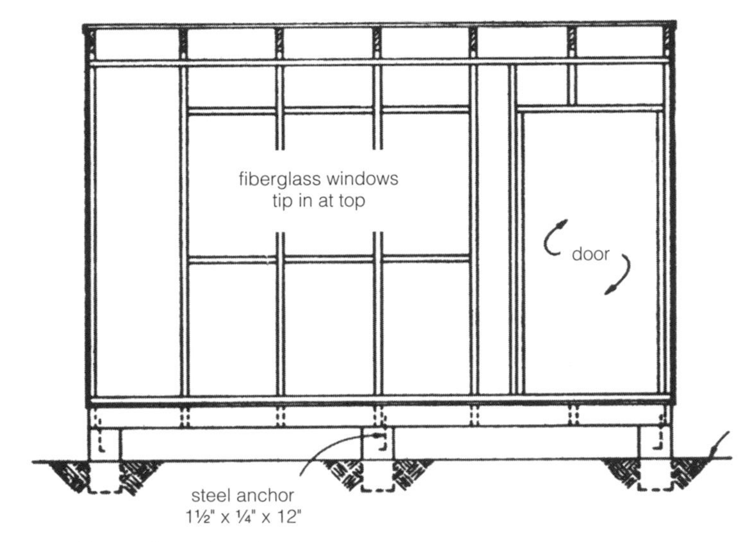

Front Elevation Framing

Framing for a 10- x 12-foot (3.1 x 3.7 m) poultry house (continued)

The completed poultry house can be easily ventilated and provides adequate floor space for the poults.

Guidelines for Providing Adequate Floor Space

It's important to provide adequate floor space for poults to avoid such problems as cannibalism. For heavy varieties, provide 1 square foot (0.09 sq m) of floor space per poult up to 6 weeks of age. From 6 to 12 weeks, increase the floor space to 2 square feet (0.18 sq m) per poult; from 12 to 16 weeks, allow a minimum of 3 square feet (0.27 sq m). Mixed sexes grown in confinement need 4 square feet (0.37 sq m) of floor space per bird from 16 weeks to market. If the flock is all toms, provide 4½ to 5 square feet (0.41–0.46 sq m) of floor space; if it is all hens, 3 square feet (0.27 sq m) is adequate. For light-type turkeys, floor space requirements may be reduced slightly.

Insulation

The amount of insulation required in the building depends on the time of the year that the turkey poults are started, as well as climatic conditions in your area. A well-insulated building conserves energy, lowers brooding costs, helps keep the young turkeys warm and dry, and makes it possible to start turkey poults during any season of the year.

Lighting

Sunlight is not necessary for brooding turkeys. However, small-flock producers may use the building for the entire growth period. In this case, adequate ventilation is essential. Windows must be placed to provide cross ventilation and necessary ventilation at critical times. Windows that tilt from the top and are equipped with antidraft shields on the sides provide good ventilation. It's important to be able to regulate windows and put them in areas that prevent drafts on the young poults. One square foot (0.09 sq m) of window area per 10 feet (3 m) of floor space is normally adequate.

Equip the pen with electricity and artificial light. Young poults need intense light to enable them to find feed and water, thereby preventing starvation or dehydration. For the first 2 weeks, provide a minimum of 12- to 15 foot-candles of light at the poult level. Bright light should be used 24 hours a day for the first 3 days. A dim night-light is usually provided thereafter to prevent piling of the confined birds. Depending on environmental conditions, brooding of the poults is usually completed after 5 or 6 weeks. They are then ready for their permanent growing quarters.

Thoroughly clean and disinfect the area to be used for brooding poults. Many good disinfectants are available from agriculture-supply houses. Whatever disinfectant is used, follow the directions on the container. Some materials can cause disinfectant injury to feet or eyes, which may damage the poults severely.

Management Systems

Several methods, or management systems, are suitable for raising turkeys.

- ***Porch-rearing*** may be necessary for birds being raised on a range, because those that have never been outside the brooder house may not seek shade from sunlight or shelter from the rain; thus, producers frequently build sunporches attached to the brooder house.
- ***Range-rearing*** is an excellent opportunity to reduce feed costs; the range may be fenced in, with range shelters and roost provided.

- ***Confinement-rearing,*** which is raising turkeys in an enclosed poultry house, is frequently used by the small-flock grower; if predators or adverse weather conditions are an issue or range or yard area is limited, confinement-rearing is necessary.
- When possible, many small-flock owners provide a ***yard*** for their birds within a fenced-in area, using the brooder house as a shelter.

The management system of choice depends on personal preference, availability of adequate housing space or range area, and possibly the type of market you wish to target. If you are developing a free-range product, the birds must have access to an outside yard or range area. It is safer to grow the poults in confinement. However, if building space is not adequate to contain growing birds to market age, other arrangements must be made.

For some, ensuring adequate space means adding a porch to the brooder house. A porch has the advantage of keeping the birds out of their droppings, thereby reducing the potential for disease problems. On the other hand, foot and leg problems may be more of a difficulty with the use of porches, especially if you are rearing heavy birds.

You can use yards to good advantage to grow more birds in a given housing area, although keeping the yard clean may be a problem. Dirty, overused yards can easily lead to health problems. However, the brooding shed can be constructed to be portable; this allows for rotation of several yards. Another management system is the use of paved, gravel, or stone-surfaced yards adjacent to the brooder or growing house. These types of yards may be easier to keep clean but can lead to leg or foot problems as the birds reach heavier body weights.

Sunporches were once very popular with turkey growers and are still used by some. These porches are usually attached to the brooder house or shelter. The floor of the porch is made of either slats or wire. The porch is elevated to provide space underneath for accumulation of droppings and easy access for cleaning. The porches are fenced in on the top and sides. Coarse mesh wire is used for the top where snow loading may be a problem. Porches are sometimes used to help acclimatize the birds to changes in weather conditions or extremes of temperature prior to going on range. Predators might also be discouraged when birds are confined to a porch.

Homemade sunporch (for detailed plan, see page 50)

Having a wire floor and, to a lesser extent, a wood slat floor on a porch may lead to foot and leg problems for heavy birds. However, such floors may be worth the risk for small-flock producers because the birds are kept away from their feces. This can be a big advantage in preventing a number of turkey diseases. In addition, newer flooring material, such as rubber-coated, heavy-wire, grate-type flooring, might be available to small-flock producers. This material is easier to clean and disinfect than wood and is more gentle on the turkeys' legs and feet than is wire. A clean yard or small range used in combination with a porch may benefit both the producer and the birds. The birds could be run on clean yards or range during good weather or for short periods and kept in the porch and brooder area during most of the growing season and especially during cold, damp weather.

Producers of large flocks frequently use environmentally controlled houses. These houses may be windowless or have large windows along each side covered by a white curtain that can be opened or closed as needed. These buildings are usually well insulated to achieve maximum heat efficiency during the brooding period and to provide comfortable, well-ventilated conditions for the birds. For more detailed information on housing and management systems, see chapter 3.

Range-Rearing: Pros and Cons

Range-rearing has advantages and disadvantages. It requires good fencing and shelters, feeders, and waterers that can be moved frequently. Good range can reduce feed costs by saving some feed, and many consumers will pay a premium for free-range birds. It is an advantage for both the turkeys and the producer to have several range areas so that birds can be rotated to a clean range. Theft, loss from predators, and other problems — such as more labor — may offset some of the benefits of range-rearing.

Equipment

The basic equipment requirements for growing a flock of turkeys include brooders, feeders, and waterers. Breeder birds require additional items, such as laying nests and egg-handling equipment.

While turkeys require some special equipment for best rearing results, most pieces can be homemade or purchased from local feed and farm-supply outlets or by mail-order (see page 183). Nothing has to be fancy. However, it is important that feeding and watering equipment be designed to adequately service the birds with a minimum of spillage or waste.

The amount of equipment required, as well as its size, varies with the age and size of the turkeys. Sufficient feeder and water space must be provided to allow each bird equal access.

Brooding Equipment

Several types of brooders are suitable for poults. The heat source may be gas or electric. If a hover-type brooder (that is, one with a canopy over the heat source) is used, allow 12 to 13 square inches (77.4–83.9 sq cm) of hover or canopy space per poult. The brooder should be equipped with a thermometer that can be easily read. Take temperature readings at the edge of the hover approximately 2 inches (5.1 cm) above the floor. (See chapter 2 for more on brooder temperatures.) The brooder size may be adjusted to meet your needs. For example, an 18 x 18-inch (45.7 x 45.7 cm) brooder would be suitable for 25 birds.

Commercial hover-type brooder

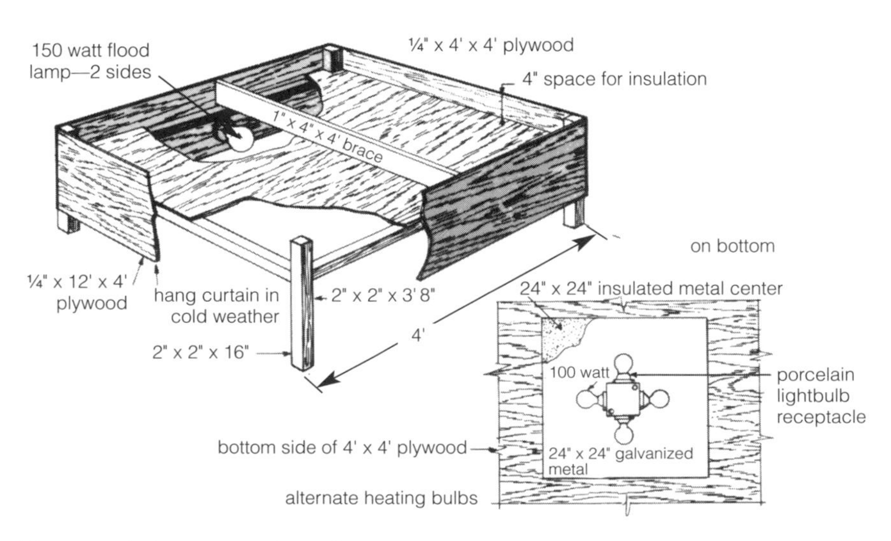

Homemade hover-type brooder

Infrared brooder lamp

Infrared brooders are satisfactory for small numbers of poults. Provide two or three 250-watt bulbs per 100 poults. Even though one lamp may be adequate for the number of poults started, an additional bulb is recommended as a safety factor in case a bulb burns out. Hang infrared lamps about 18 inches (45.7 cm) from the surface of the litter at the start. After the first week, raise them 2 inches (5.1 cm) each week until they reach a height of 24 inches (61 cm) above the litter. The room temperature *outside* the hover or brooder area should be approximately 70°F (21°C) for maximum poult comfort.

Where available, battery brooders can be used for the first 7 to 10 days to get the poults started. Allow approximately 25 square inches (161.3 sq cm) of space per poult in the battery. After the poults are removed from the battery and placed on the floor, watch them carefully to make sure they learn to use the feeders and waterers. Also, poults that have been reared in battery brooders and then put onto floors with pine shavings as a bedding material will probably consume some of the shavings. This could lead to crop impaction. Providing a little grit in the battery brooder may help to develop the gizzard and prevent crop impaction.

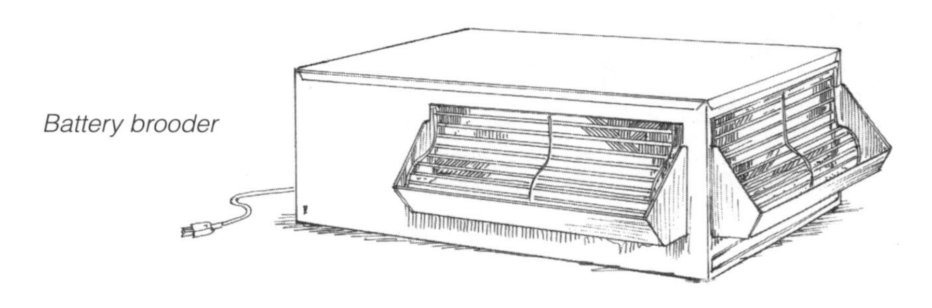
Battery brooder

Feeding Equipment

To quickly get the poults to start eating, place their first feed on egg-filler flats, chick box lids, paper plates, small plastic trays, or box covers. When the poults arrive, place them in the feed container. When one or more of the birds starts to peck at the feed, it will attract the others.

When box tops or egg-filler flats are used as the early feeders, spread them around in the brooding area among the regular-type feeders. After 3 days, gradually move these feeders toward the regular feeders. Usually, at 7 to 10 days, the early feeders can be removed and the poults will use the regular feeders. You may want to lay paper underneath the feeders for the first few days to prevent litter eating. If you choose not to cover the litter with paper, be sure *not* to fill the feeders so full that the feed overflows onto the litter, because this may lead to the practice of litter eating. Use of smooth-surfaced paper is not recommended because slippery surfaces, used for prolonged

Hanging, tube-type feeders and waterers are excellent for poults and for turkeys.

periods, can cause foot and leg problems in young poults. If available, use paper with a rough surface.

It is also a good idea to dip each poult's beak into a water trough to make sure the bird experiences where to find water. Dehydration is a common malaise of poults.

After 2 or 3 days, the young poults should be acclimated to the location of the feeders and waterers and fully utilizing them. At that point, the paper can be removed and the poults allowed free access to the litter. Careful observation of the birds and how well they use the feeders and waterers will give you a sense as to when it is appropriate to remove the paper.

Shallow pans or egg cartons are fine for beginning feeders.

Techniques to Get Poults Feeding Quickly

It is important to get poults started early on feed and water. If they don't find the feed and water easily, starvation or dehydration can occur. When the birds are small, chick-size feeders are used at first. Bright-colored marbles or other colored objects are sometimes placed in the feed and water containers. These help attract the poults to the feed and water. Oatmeal, other cracked grain, or fine granite grit sprinkled lightly over the feed once or twice a day for the first 3 days may also help to get them eating.

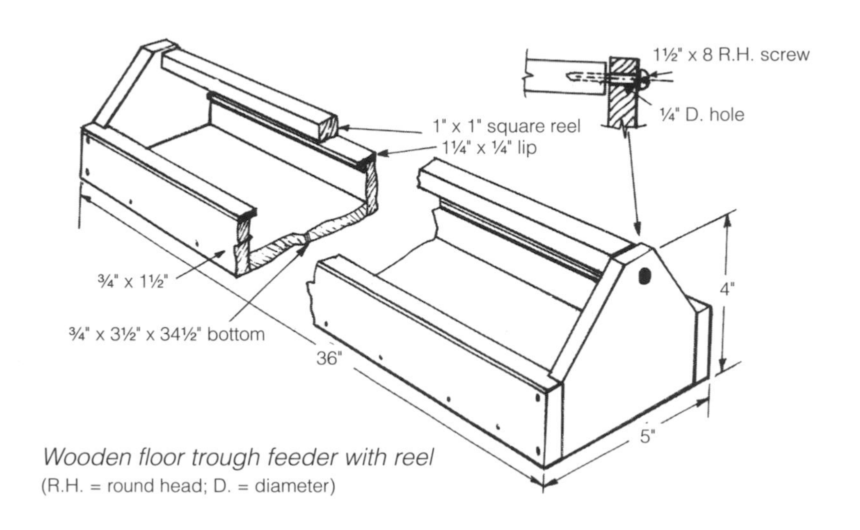

Wooden floor trough feeder with reel
(R.H. = round head; D. = diameter)

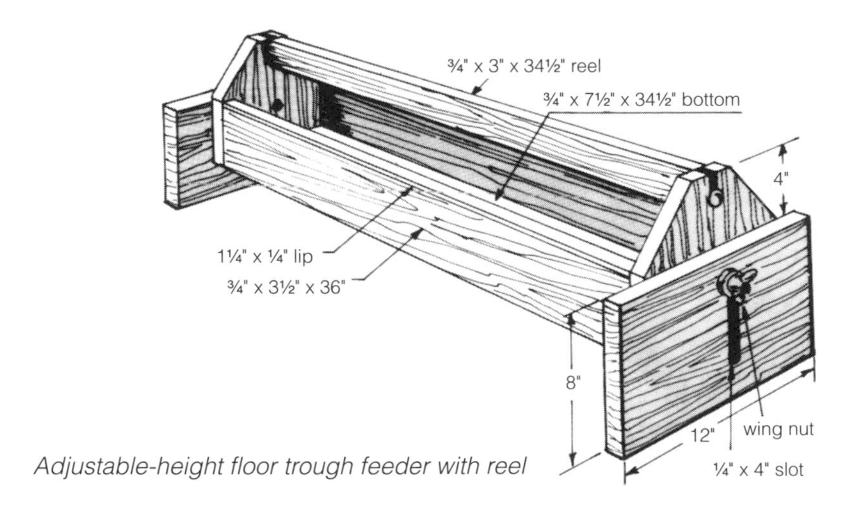

Adjustable-height floor trough feeder with reel

Determining Feeder Size

From 7 days to 3 to 6 weeks of age, use small feeders. Provide 2 linear inches (5.1 cm) of feeder space per bird. From 3 weeks to market age, the poults should have access to larger feeders about 4 inches (10.2 cm) deep, with 3 linear inches (7.6 cm) of feeder space per bird. Hanging, tube-type feeders are excellent for turkey poults.

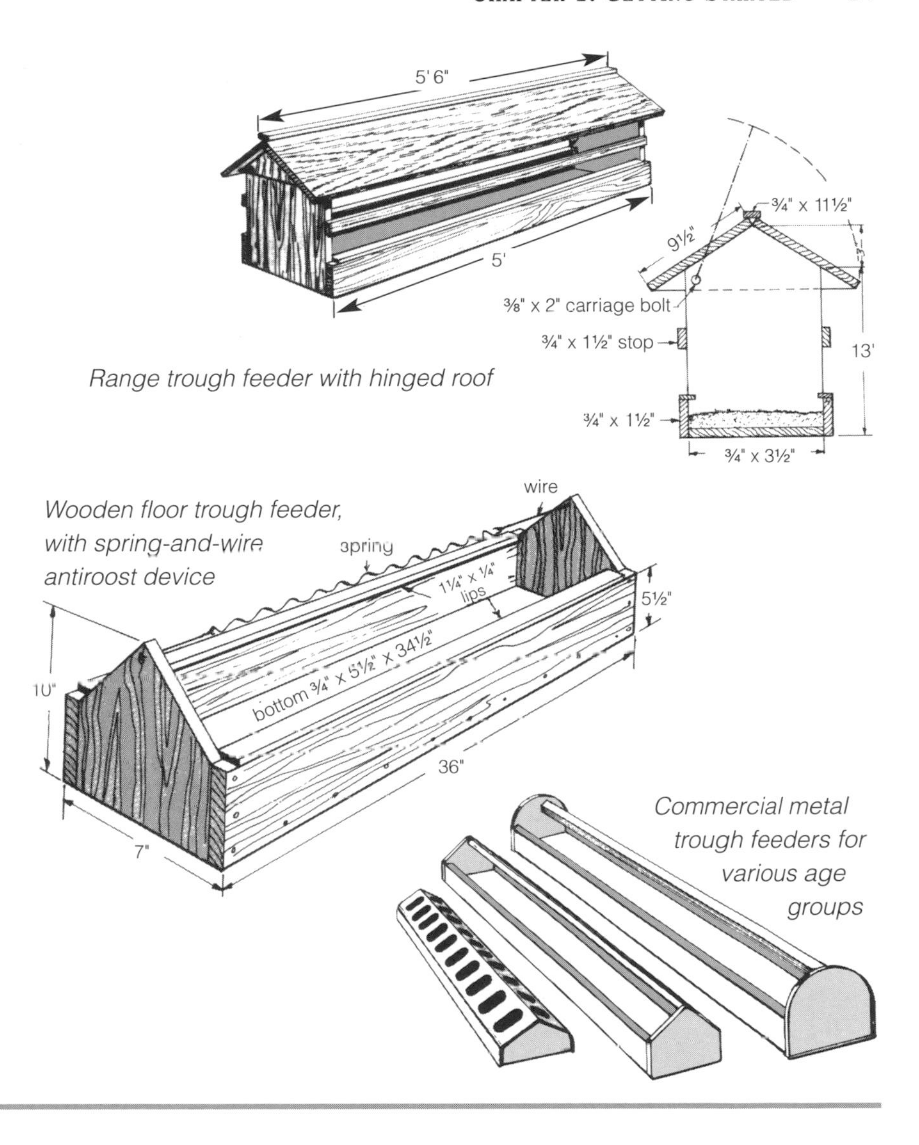

Range trough feeder with hinged roof

Wooden floor trough feeder, with spring-and-wire antiroost device

Commercial metal trough feeders for various age groups

The amount of available tube-type feeder space can be determined by multiplying the diameter of the feeder pan by 3. When figuring feeder space, remember to multiply the hopper length by 2 if the poults are able to use both sides of the feed hopper. Thus, a 4-foot (10.2 cm) trough feeder actually provides 8 linear feet (2.4 m) of feeder space. There are several types of feeders that can be purchased or built at home.

Watering Equipment

Poults are usually started on either glass or plastic fountain-type waterers or automatic waterers. From 1 day to 3 weeks of age, they should have access to three 1- or 2-gallon (3.8 or 7.6 L) fountains per 100 poults. From 3 weeks to market age, they should have two 5-gallon (19.2 L) fountains per 100 poults or one 4-foot (1.2 m) automatic waterer or two small bell-type waterers. For smaller flocks, adjust the number and size of waterers as necessary. Note that providing adequate water space and water is imperative for good turkey production in any season but is especially important during warm and hot weather. A 20-degree F (11-degree C) increase in ambient temperature can double water consumption. Water temperature can also affect consumption. Providing water that is cooler than ambient temperature during hot weather is a good idea because it encourages the birds to drink and helps alleviate heat stress. It is also imperative to keep water troughs clean. There is no magic number of times per week that waterers need to be cleaned — it takes as many times as it takes! Automatic waterers and water tanks can be obtained from poultry-, livestock-, or agriculture-supply centers.

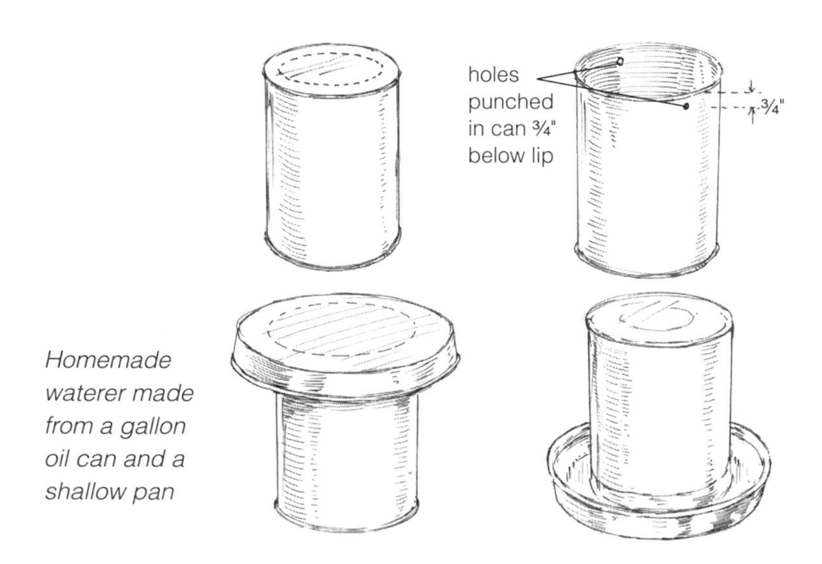

Homemade waterer made from a gallon oil can and a shallow pan

Various types of waterers

poult waterer

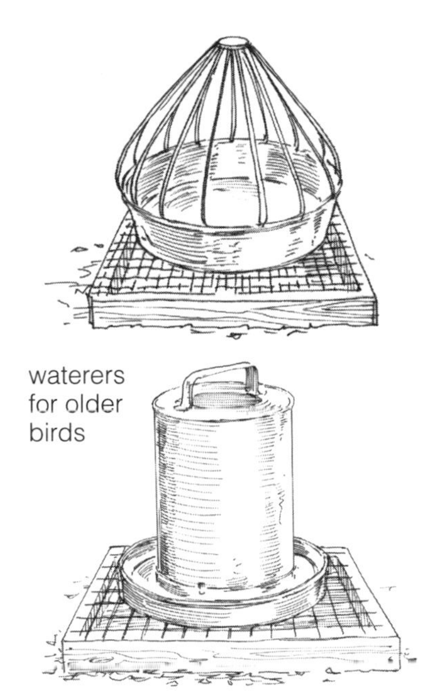

waterers for older birds

Miscellaneous Considerations for Feeding and Watering

Change equipment, both feeders and waterers, gradually to avoid discouraging feed and water consumption. For older birds, waterers can be placed on wire platforms. (The dimensions of the wire platform depend on the size and type of waterer used.) This helps prevent litter from fouling the waterers, keeps the poults out of the wet litter that frequently surrounds the waterers, and keeps thc litter in better condition.

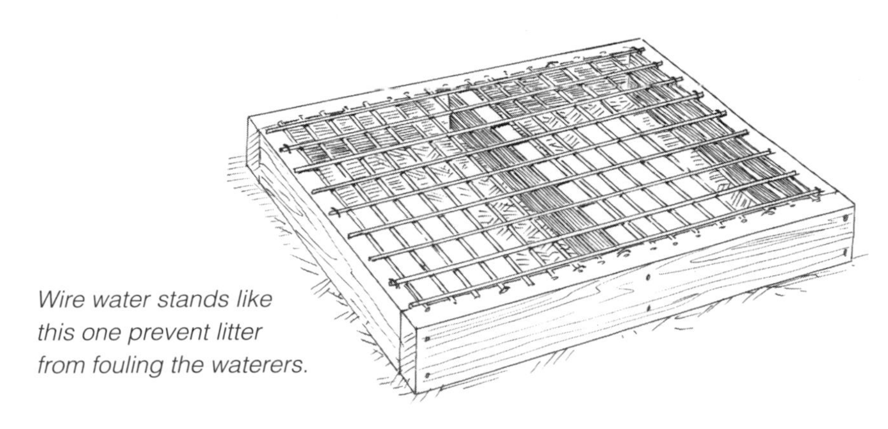

Wire water stands like this one prevent litter from fouling the waterers.

Alternatively, move the waterers often to avoid wet litter buildup. Once the waterer is moved, the litter beneath the old location can be turned to encourage drying, or it can be removed as needed.

Roosts

Roosts, or perches, are not used frequently for turkeys during the brooding period and are not necessary, though they do help prevent piling at night. Sometimes flashing lights, sudden noises, or rodents running across the floor can startle the poults. They may pile into a corner and cause injury or smothering. Normally, the birds begin to use roosts at 4 or 5 weeks of age. If roosts are used, they may be either the stepladder type or merely flat frames with perches on top. Following are some recommendations for roosts in the brooder house:

- Make roosts out of round poles that are 2 inches (5.1 cm) in diameter or 2 x 2-inch or 2 x 3-inch (5.1 x 5.1 cm or 5.1 x 7.6 cm) material.
- Place them 12 to 15 inches (30.5–38.1 cm) above the floor.
- Allow 6 linear inches (15.2 cm) of roost space by the end of the brooding period.
- Screen the sides and ends of roost pits to keep the poults away from the droppings.
- If they are placed in a house or shelter, the roosts may be slanted to conserve space.

Where birds are grown on range, roosts are used quite frequently for young and smaller birds. Older birds, especially market-age toms, will probably not use roosts. Remove the roosts if space is limited and the birds have stopped using them. Roosts for range birds can be constructed in the following manner:

- Use 2-inch (5.1 cm) poles or 2×4 material laid flat with rounded edges.
- Space the roosts 24 inches (61.6 cm) apart, 15 to 30 inches (38.1–76.2 cm) off the ground.
- Build all roosts on the same level.
- Build outdoor roosts out of fairly heavy material to prevent breaking when the weight of the birds is concentrated in a small area.
- Allow 10 to 15 inches (25.4–38.1 cm) of perch space per bird for large-type birds and 10 to 12 inches (25.4–30.5 cm) for small-type turkeys up to maturity.

Homebuilt roost for range birds

Litter Conditions

When turkeys are grown completely in confinement houses, roosts are normally not used. The birds bed down in the litter on the floor and, therefore, litter conditions must be sanitary to prevent such problems as breast blisters, soiled and matted feathers, and off-colored skin blemishes on the breast. More important, good litter conditions improve sanitation and prevent disease.

To maintain optimum litter conditions, remove wet or caked litter and replace with clean, dry material. Good pen ventilation helps remove excess moisture and keeps litter dry, as does rigorous management of water and waterers.

Points to Remember

- The turkey is a North American bird that was first used and domesticated by Native Americans.
- The turkey is a major U.S. agriculture product but can be grown in small flocks by anyone with a little space and time and a love of raising animals.
- If you are raising birds for profit, raise only what you can use or sell.
- Detailed management is the key to raising turkeys successfully.

TWO

Brooding the Poults

Brooding young turkey poults is one of the most enjoyable and important phases of turkey raising. During this period, the young poults grow rapidly beneath a warm brooder in a confined area. A *brooder* is a heating unit that provides the warmth necessary to help the poult maintain its body temperature until it is old enough to do so on its own. Brooders can be gas or electric. Several types can be purchased from commercial or mail-order poultry-supply companies, agriculture-supply companies, or feed- or farm stores.

One of the most important factors in brooding is to start off with good-quality poults. All poults are not equal in their health, liveliness, or general ability to live and grow. Good-quality poults are a joy to raise. Poor-quality poults have decreased chances for survival and may experience difficulties for their entire rearing period. Another critical factor is providing proper care to young poults, especially during the first 2 to 3 weeks. Providing good-quality feed and adequate, clean fresh water is also extremely important.

Turkeys have been improved through careful breeding to the point that remarkable performance is now possible. If you start with good-quality poults and provide adequate housing, feed, and water, you will have a strong foundation for a turkey enterprise. However, to be successful you must adopt a sound management philosophy that takes into account the necessity of providing essential nutrition and using disease-control programs. Careful, vigilant management is key.

Preparing the Brooder House

There are certain precautions you should take prior to placing the young poults in the brooding area. If the brooder house has been used previously for chickens or turkeys, it is extremely important to clean and disinfect the house and equipment before the young poults are set down. This means completely removing all litter and any caked material adhering to walls, floors, or equipment. Wash the floors, walls, ceilings, and equipment thoroughly. The importance of cleaning the equipment and facilities cannot be overstated. Dirt cannot effectively be disinfected.

Disinfecting Brooding Areas

After *thorough* cleaning, disinfect the building and equipment with cresylic acid or one of the phenol or quaternary ammonium compounds available from farm-supply houses. Follow the label directions carefully to avoid disinfectant injury to the poults or to the person applying the disinfectant. After cleaning and disinfecting the equipment and facilities, let the brooder house dry and air out for 3 to 4 weeks prior to placing poults back in the house.

Litter Material

After the building is thoroughly cleaned, disinfected, and dry, place about 4 inches (10.2 cm) of litter material on the floor. Good litter is absorbent, lightweight, of medium particle size, and a good insulator.

The following materials make good litter or bedding:

- Softwood shavings (such as pine)
- Rice hulls
- Ground corncobs
- Finely chopped straw

Hardwood shavings and peanut hulls do not make desirable litter because they tend to become moldy and contaminated with other pathogens. Sawdust is also not a good litter material because the poults are more likely to eat it, and it provides no nutrition and may lead to crop and gizzard impaction. In addition, small-particle litter leads to an increased amount of caked litter around waterers and feeders. This, in turn, can lead to increased leg and foot problems if the cake is not removed.

On the other hand, large-particle litter, such as wood chips or coarse straw, does not absorb moisture very well and can also lead to foot problems. The commercial poultry industry has experimented with many alternative litter materials. However, clean, dry pine shavings are still the preferred material and serve the small-flock producer well. If pine shavings are difficult to obtain, you can try dry, clean straw. Straw is abundant in many areas and should be easy to obtain. It is not very absorbent but can work for small flocks when the caretaker is able to change the straw as often as necessary. In addition, the absorbency of straw can be increased if it is chopped in about 2-inch (5.1 cm) pieces. This can be accomplished with a dairy tub chopper.

Litter may be covered or uncovered. Some producers cover the litter with paper for 3 to 7 days after the poults' arrival to prevent them from eating it. As mentioned earlier, if the litter is covered, use rough paper to prevent foot and leg problems. Most people do not cover the litter. Distribute the litter very evenly over the floor, and make sure it's dry and free of mold and dust. Very coarse litter can also contribute to leg disorders, while fine materials can be too dusty. For small brooding rooms, it may be a good idea to round the corners of the brooder house with small-mesh wire or a solid material, such as cardboard or brooder ring paper, to prevent piling in the corners while the birds are young.

Benefits of Litter

Litter material serves many important functions. It insulates the floor and helps to conserve heat while providing increased comfort for the birds. It also dilutes and absorbs moisture from fecal matter, condensation from bird respiration, and water spilled from drinking fountains. Providing a soft, spongy surface on which turkeys can rest helps to prevent breast blisters and *buttons* (fecal matter on feathers), and helps to satisfy birds' dusting and scratching instincts as well.

Brooder Guards

Use brooder guards or brooder rings to confine the birds to the heat source and the feeding and watering equipment until they become accustomed to their environment. Solid brooder-guard material also prevents drafts on the poults. The brooder guard should be 14 to 18 inches (35.6–45.7 cm) high. It can be made of corrugated cardboard, which is available in rolls that are several feet long. This is the most convenient and easiest material with which to work. In addition, it can be thrown away after use, which eliminates the risk of disease for the next flock of poults. It may be tempered Masonite or prefabricated panels that clip together to form a ring around the brooding area. For warm-weather brooding or houses in which drafts are not a problem, the brooder guard can be made from poultry wire secured to frames.

Managing the Brooder Guard

The design of the house and climatic or seasonal conditions determine appropriate management of the brooder guard. When a noninsulated house is used during fairly cool weather (50°F [10°C] or below), an 18-inch (45.7 cm) brooder guard for each stove or heat source is recommended. For warm-weather brooding, a 12- or 14-inch (30.5 or 35.6 cm) brooder guard is satisfactory. At first, place the brooder guard 2 to 3 feet (0.6–0.9 m) from the edge of the heat source and then gradually move it out to a distance of 3 or 4 feet (0.9 or 1.2 m) from the edge of the heat source. Remove the guard on the tenth day, or when the poults start to hop over it. Set up the brooder guard carefully so that there is 6 to 12 inches (15.2–30.5 cm) of space between feeders or waterers and the brooder guard; this permits traffic around the ends of the feeders.

A brooder guard placed around the brooder area helps birds become accustomed to their environment. Remember, young poults soon learn to hop and fly, so place the brooder in a confined area that is well protected from cats, dogs, and other predators. Ideally, the brooding area should be a separate shed or building.

Preparing a Brooding Area

Set up and establish the brooding area about 48 hours before the poults are due to arrive. This is especially important during cool or cold weather and even at times when it is cold only at night. The floor and litter material must have time to warm up. A cool floor and litter can act as a heat sink that pulls warmth from the poults even though the brooder is operating properly. Fill the feeders and waterers and have everything ready so the poults can be removed from the containers as soon as they arrive. Again, it is important that they be put onto feed and water as quickly as possible. If you have only a small number of poults, remove each one from the shipping box and dip its beak first in water and then in the feed. This encourages the poults to drink and eat. Time spent making sure all of the poults have been introduced to water helps ensure a successful start.

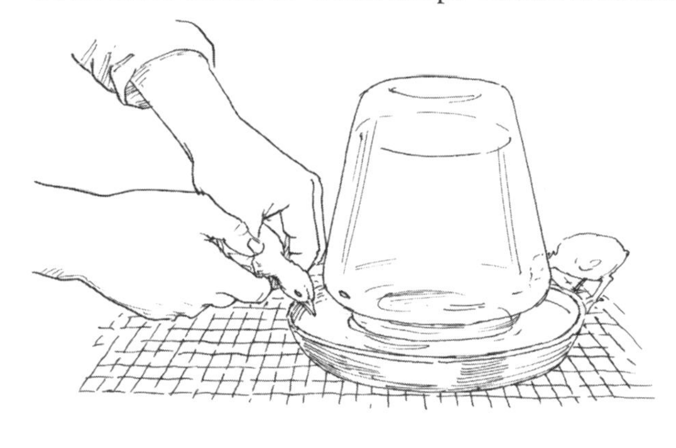

Dipping a poult's beak first in water and then in feed encourages the poult to drink and eat.

Maintaining Appropriate Temperature

If hover-type brooders are used, the temperature should be approximately 95°F (35°C) the first week. Take this temperature reading at the edge of the hover approximately 2 inches (5.1 cm) above the litter, or at the height of the poult's back. Be sure to check the accuracy of the thermometer before the poults arrive. Reduce the hover temperature approximately 5°F (2.8°C) weekly until it registers 65° to 70°F (18°–21°C) or is equivalent to the prevailing nighttime environmental temperature — whichever is greater. During the first several weeks of brooding, room temperature should be approximately 70°F (21°C). This helps to prevent drafts on the poults and prevents wide temperature fluctuations in the brooder room.

If the weather is warm during the brooding period, the heat can be turned down at the beginning of the second week. However, heat should be maintained during the evening hours for a longer period. Normally, little or no heat is required after the sixth week, depending on the time of year, weather conditions,

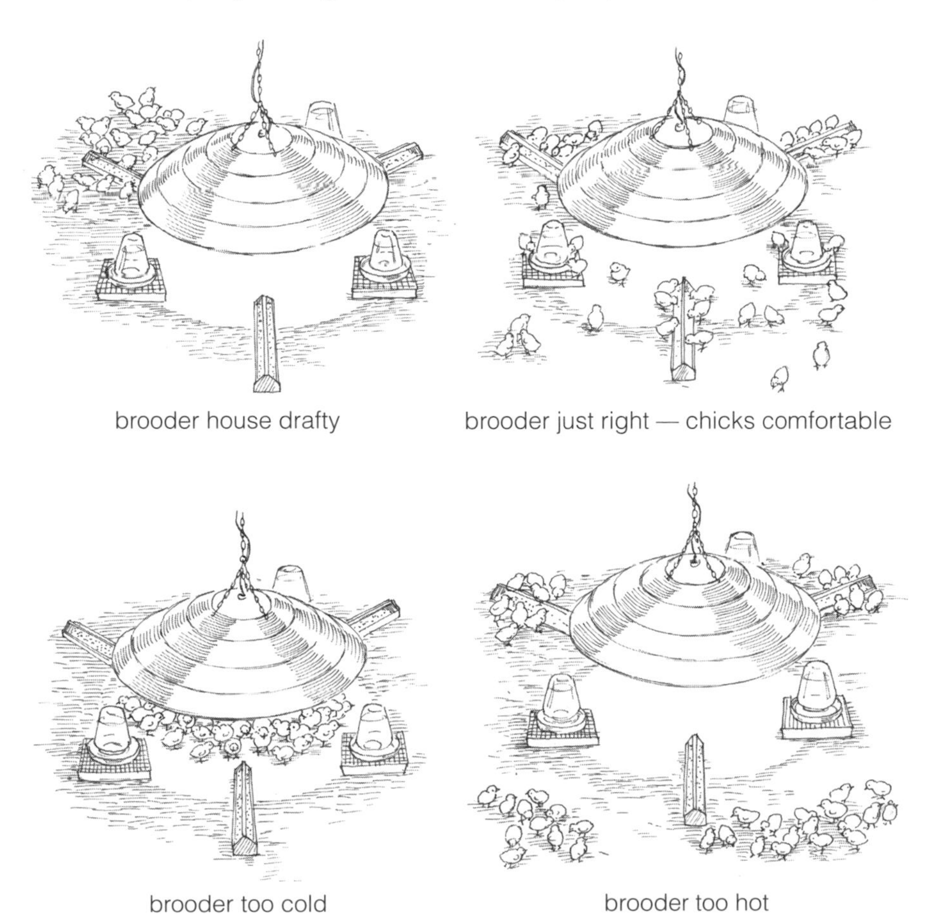

Watch the poults to determine whether the brooder is too cold or too hot.

and housing. After the first week or so, experienced poultry producers can watch the poults and tell whether they are comfortable. It may be necessary to turn on the heat on cool evenings to keep the birds warm enough.

Lighting

Give the poults high light intensity for the first 1 to 2 weeks of brooding. This ensures a good start by increasing their activity and helping to prevent starvation. Infrared brooders provide adequate light intensity for the poults. If hover-type brooders are used, artificial light combined with the attraction light should provide a minimum of 12 foot-candles of light at the feeder and water level. After the first 2 weeks, about 1 foot-candle is adequate, and this lower intensity helps reduce nervousness and flightiness in the flock.

Floor Space

Once poults are no longer being confined by the brooder ring, provide 1 square foot (0.09 sq m) of floor space for each poult up to 6 weeks of age. When they are 6 to 12 weeks old, increase the floor space allowance to 2 square feet (0.18 sq m) per poult. From 12 to 16 weeks, the minimum allowance is 3 square feet (0.28 sq m). When both sexes are kept together in confinement during the entire growing period, provide 4 square feet (0.37 sq m) per bird. If the flock is all toms, 5 square feet (0.47 sq m) of floor space is desirable; if it is all hens, 3 square feet (0.28 sq m) is adequate. For light-type turkeys, the floor space requirements can be reduced. It's important to observe space requirements to avoid cannibalism and feather pulling and to make sure birds get adequate feed and water. As you gain experience in turkey production, floor space can be adjusted as needed or desired.

Brooding Sanitation

Following these steps will ensure adequate brooding sanitation:

- Clean all waterers with a brush and disinfect them daily. Quaternary ammonia and organic iodines are suitable disinfectants. Do not spill water in brooder areas.
- Make equipment, feed, and management changes gradually.
- Move portable feeders and waterers daily to help eliminate damp or wet litter. After approximately 1 week, keep feeders and waterers adjusted to be level with the poults' backs.
- Remove or stir damp or caked litter daily to prevent mold buildup.
- Maintain desired house temperature.
- Monitor ventilation as needed to provide some fresh air and to remove dust, moisture, and ammonia.
- Remove dead poults immediately, and keep a record of each occurrence. Data should include the date, number, and any observations of the poult at time of death.
- Use a dip pan at entrance for disinfecting the caretaker's shoes or boots. ▶

- No visitors!
- Keep other animals out of the brooder area and the entire growing area if possible. Wild animals can transmit diseases.
- Practice good animal husbandry. The turkey poult, as does any young animal, requires careful attention.

Close observation can teach the caretaker a lot about turkey poults, especially about their comfort level. Most poult mortality and growth problems can be traced to poor husbandry practices. This includes lack of attention to detail, poor sanitation, infrequent trips to visit the poults, faulty equipment (such as brooders that shut off), and inadequate ventilation.

THREE

Managing Turkeys Effectively

There is absolutely no substitute for good animal husbandry. Good husbandry can be defined simply as providing for the needs of the birds as those needs arise. *Management* and *husbandry* are used interchangeably to the point where you seldom hear the term *husbandry* any more. However, I prefer to define *husbandry* as management with sincere concern for the turkey's well-being added in. For the purpose of the following discussion, *husbandry* and *management* will be used interchangeably.

It is not difficult to raise turkeys as long as you start with good poults, feed and care for them well, and avoid major disease outbreaks. Good management is an important factor — without it, optimal results will not be realized. Certain tasks and practices are peculiar to the type of turkey being raised or to the management systems used. A management system is a method of raising and housing a flock. Regardless of the system used, however, attention to detail is a crucial component of any livestock management system.

There are many different management options, but the type of management system you use will depend on factors such as location, size of proposed enterprise, amount of quality land available, and water supply.

General Management Recommendations

With good management, you should be able to raise to maturity 85 to 90 percent of the turkeys started. With the high costs of poults and feed, mortality can become expensive, especially when the birds are lost during the latter part of the growing period. The following management techniques can help.

Keys to Good Management

You can achieve good husbandry simply by following these practices:

- Keep young poults isolated from older turkeys, chickens, and other poultry. Ideally, no other birds should be on the same farm where turkeys are raised.
- Take care to avoid tracking disease organisms from older stock to young stock or from other birds to the turkeys.
- Follow a good control program for mice and rats. These rodents carry disease and are capable of consuming large quantities of feed. Rats can kill young poults, too.
- If abnormal losses or signs of disease occur, immediately take the birds to a diagnostic laboratory for diagnosis. (See Diagnostic Laboratories by State, page 166.)
- Watch consumption on a daily basis. One of the first symptoms of a disease problem is a reduction in feed and water consumption.
- Look for disease problems in your birds if sudden changes occur that cannot be traced to temperature or to other stresses.

Depending on the diseases present in your given area, it may be necessary to vaccinate the poults against Newcastle disease, fowl pox, erysipelas, or fowl cholera. To plan a vaccination program for your flock, check with the poultry diagnosticians at your state animal pathology laboratory, your county agricultural agent, your Extension poultry specialist, your state poultry federation, or some other knowledgeable person. Frequently, well-managed small flocks are not vaccinated and have no disease problems, especially when mixing different species of poultry is avoided.

Antibiotics are now being added to feed to prevent numerous diseases. Antibiotics and other drugs are valuable in preventing and treating diseases, but don't use them as a substitute for good management. Small-flock producers should concentrate their efforts on providing clean rearing facilities and not on antibiotics for preventing diseases. If there is a disease outbreak, antibiotics are more effective if the disease agents have not already been exposed to antibiotics. Dietary coccidiostats are recommended, however, to prevent outbreaks of coccidiosis. For more information on turkey diseases and their prevention, see chapter 5.

A good manager checks the turkeys frequently.

Management Systems

Several types of management systems may be used successfully to grow turkeys. The majority of commercial producers now grow their turkeys in confinement. The poults are usually brooded in the turkey house and stay in the house until marketed. They are frequently started in a small portion of the house and, as the birds become larger and need more space, gradually permitted to use the whole house.

Porch-rearing was once a popular method. Porches are attached to the brooding facility. When the birds reach a few weeks of age, they are allowed to go out on the porch area for a few weeks before being put out on range. Some are raised to market age on the porches. This is still a viable way of growing small flocks of turkeys.

Another method used, where suitable land is available, is range-rearing. The birds are started in a brooding facility and, at roughly 6 to 8 weeks of age, are put out on range for the remainder of the growing period, depending on the weather.

There are also variations of some of these methods, all of which may be used successfully by small-flock producers. And remember that none of these methods is mutually exclusive. You can use any one method as described, or you can borrow from all of them to create your own unique system.

Porch-Rearing

As mentioned, turkeys that have never been outside the brooder house may not seek shade from sunlight or shelter from the rain when placed on range. For this reason, producers who are going to range their turkeys frequently put their birds on sunporches attached to the brooder house. When the weather is warm, the young turkey poults can leave the brooder house and go out on the porches as early as 3 weeks of age. Usually, the porches are covered with fine-mesh woven wire on the sides and top to prevent the poults from getting out. The floors may be either slat or wire. Either one works well if the porches will be used for just a few weeks before the birds go on range. However, if the birds are to be raised on porches up to market age, wire floors are not satisfactory, particularly for heavier turkeys. Some birds tend to develop foot and leg problems as well as breast blisters or sores.

Smaller varieties of turkeys and those to be dressed at an early age for broilers or fryers do quite well on wire floors. Grow larger varieties for heavy roasters on porches with slat floors.

Locating the feed and watering equipment so that they can be serviced from outside the porch greatly simplifies the chores.

Pros and Cons of Porch-Rearing

Many small turkey flocks are successfully grown on porches. As with confinement-rearing (see page 47), birds grown on porches are not as likely to be attacked by predators as are those grown on range or in a yard. They are also a lot less likely to develop disease problems, particularly litter- or soilborne diseases. On the other hand, if adequate space is not available and the birds have not had their beaks trimmed, feather pulling and cannibalism tend to be more common in porch-reared birds.

Range-Rearing

Range-rearing offers an opportunity to reduce the cost of growing turkeys. This is especially true if the diet can be supplemented with homegrown grains. Turkeys are good foragers. And if good green feed is available on the range, this means less consumption of expensive mixed feed, thereby reducing the cost of the feeding program. Building costs are much lower when birds are range-reared, but labor requirements are higher.

Portable Range Shelters

Depending on climatic conditions, some growers provide only roosts for turkeys on range. Some actually allow the turkeys to sleep on the ground. This method is more practical when the turkeys will be matured early and before the cold winter weather sets in. Portable range shelters give the turkeys much better protection during poor weather. They can be moved to new locations to provide the birds with better range conditions and prevent development of muddy spots and contaminated areas. When portable shelters are used with roosting quarters, the feeders and waterers can be moved whenever the grass is closely grazed in an area. Commercial producers sometimes provide pole buildings for shelter at night and let birds out on range during the day.

Considerations for Portable Range Shelters

Make sure there is enough space for all birds to get into the shelter at one time.

- Typical dimensions of portable shelters are 10 x 12 feet (3 m x 3.7 m) or 12 x 14 feet (3.7 m x 4.3 m), but they can be built smaller to accommodate small flocks. If built any larger, they are not as easily moved, and there is greater chance for building damage during a move.
- Portable range shelters should provide a minimum of 4 square feet (0.37 sq m) of space per large bird and 2 square feet (0.19 sq m) for small-type birds.
- A 10 x 12-foot (3 m x 3.7 m) shelter can supply roosting space for up to 60 twelve-week-old turkeys and up to thirty mature birds.

Curtains that roll down from the top might be installed to block rain from the prevailing winds on one or two sides during stormy weather. This is especially important if the weather is bad shortly after young birds are moved onto a range.

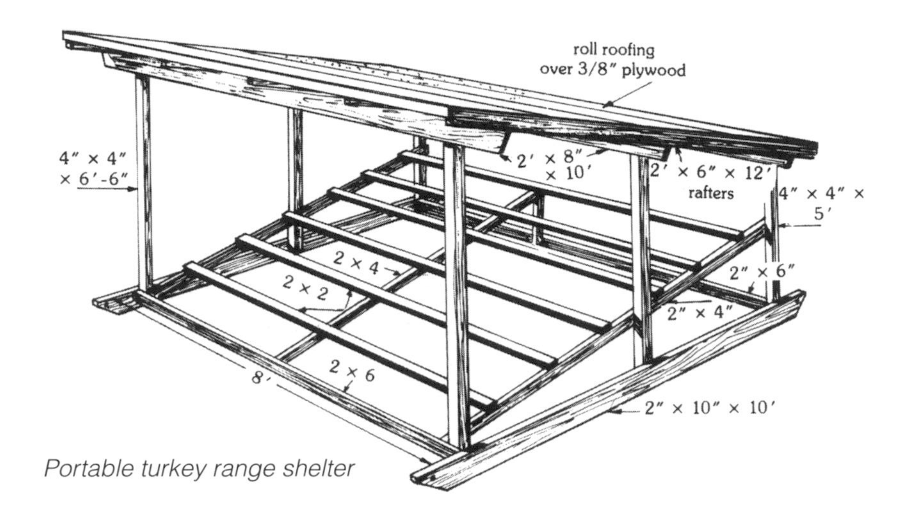

Portable turkey range shelter

Precautions and Special Considerations

Normally, May- and June-hatched poults can be put out on range by 8 weeks of age. Before putting them on range, make sure they are well feathered, especially over the hips and back. Check the forecast, and try to move them out during good weather. It is best to move the birds in the morning to give them time to adjust to their new environment before darkness.

If possible, provide a range area that has been free of turkeys for at least 1 year and preferably 2 years. You can use a temporary fence to confine the flock to a small part of the range area. Move the fence once a week or as often as the range and weather conditions indicate. Permanent fencing may also be necessary to deter predators. Wild animals or dogs can cause losses on ranges by killing or injuring birds or by causing stampedes resulting in suffocation and injuries. A 6-foot-high (1.8 m) poultry fence around the range area helps prevent these problems. Some producers add electric fencing for more protection. A single wire, 6 inches (15.2 cm) from the ground outside the permanent fence, works well.

Provide artificial shade if there is no natural shade. Several rows of corn planted along the sunny side of the range area provide good shade and some feed. If range shelters are used, move them every 7 to 14 days, depending on the weather and on the quality of the range. Move the feed and watering equipment as needed to avoid muddy and bare spots.

Range-Rearing Drawbacks

Range-rearing also has its problems. Losses can occur from soilborne diseases, adverse weather conditions, predators, and theft. Because of the potential for these problems and the additional labor required, confinement-rearing (see Confinement-Rearing, page 47) has quite rapidly replaced range-rearing in recent years.

Range Crop Choices

The range crop selected depends on the climate, the soil, and the range management. Many turkey ranges are permanently seeded. Others are part of a crop rotation plan. As part of a 3- or 4-year crop rotation, legume or grass pasture and annual range crops, such as soybeans, rape, kale, sunflowers, reed canary grass, and Sudan grass, have been used successfully. Sunflowers, reed canary grass, and Sudan grass provide green feed as well as shade. For a permanent range, alfalfa, ladino clover, bluegrass, and bromegrass are very satisfactory.

Range Feeders and Waterers

Range feeders should be waterproof and windproof so that the feed does not spoil or blow away. Place the feeders on skids or make them small enough that they can be moved by hand or with the help of a small tractor. Trough-type feeders are inexpensive and relatively easy to construct. Specialized turkey-feeding equipment can also be purchased. To minimize waste, all feeding equipment should be designed so that it can be adjusted as the birds grow; the lip of the feed hopper should be approximately in line with the bird's back. For the same reason, the feed hopper should never be more than half full. Pelleted feeds are less likely to be wasted on range. Provide at least 6 inches (15.2 cm) of feed trough per bird if the feeders are filled each day. When feeders with storage capacity are used, less space is required, and the amount of feeder space should conform to the equipment manufacturer's recommendations.

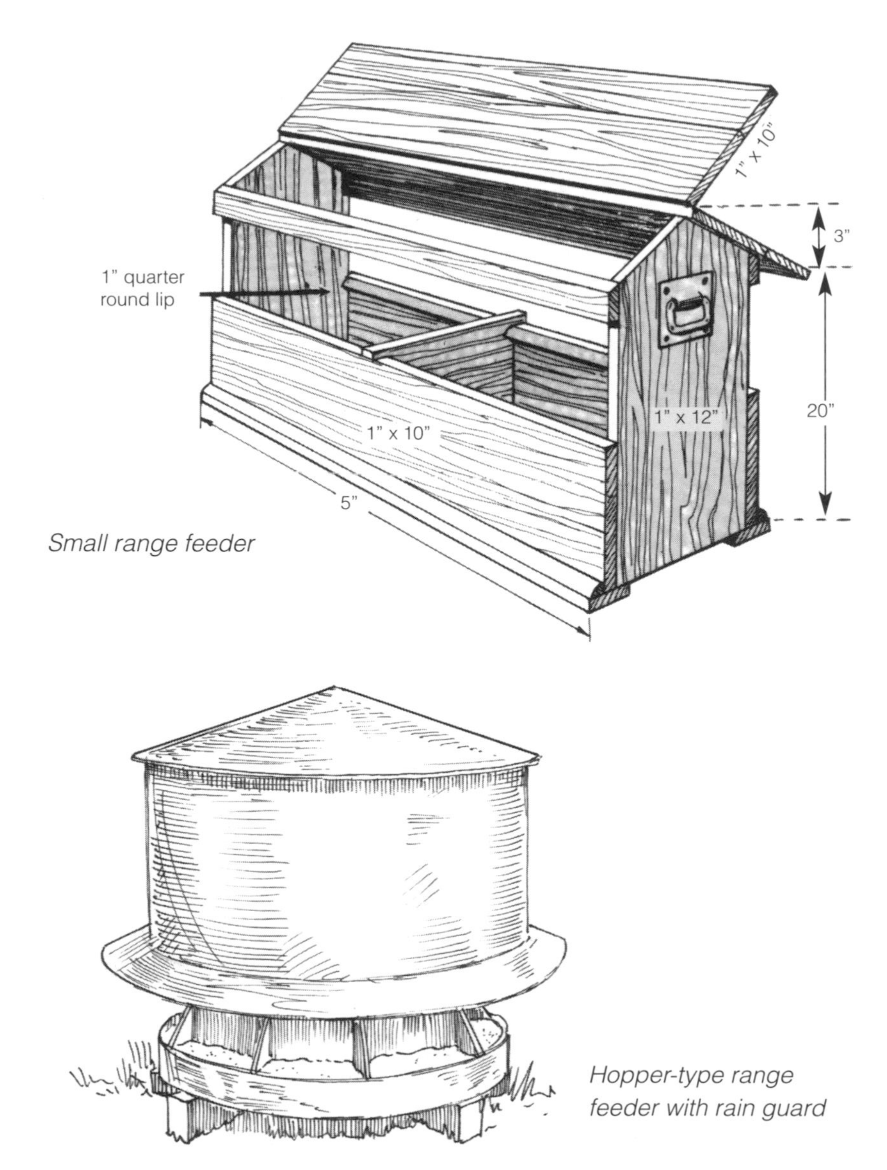

Small range feeder

Hopper-type range feeder with rain guard

Provide one 4-foot (1.2 m) automatic trough waterer, one large round waterer, or two bell-type drinkers per 100 birds. Clean the waterers daily and disinfect them weekly. Locate waterers close to the shelters. If possible, shade the waterer with portable or natural shade.

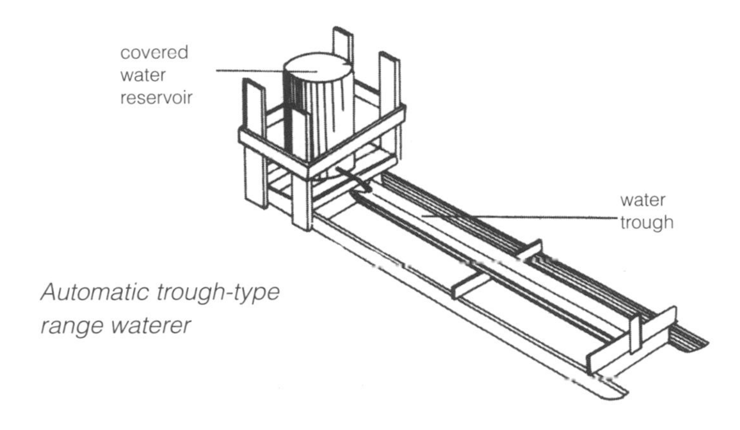

Automatic trough-type range waterer

Selecting a Range Site

With adequate land, turkeys can easily be reared on range. However, some problems can occur, including infiltration by predators such as foxes or feral dogs or cats, outbreaks of disease and exposure to bad weather such as rain, sleet, snow, and severe heat or cold.

Proper site selection and management can help prevent some of these problems or at least help in the control and management of problems when they do occur. Other factors to consider include the type of soil and soil drainage, the type and quality of vegetation on the proposed site, whether shade is available or can be provided, and how best to choose a site that will discourage losses from theft or predators.

Soil type and drainage. Poorly drained soil does not make good range for turkeys. Stagnant surface water can be a source of disease. Therefore, quick and complete drainage of turkey ranges is essential regardless of location. Range site selection in certain geographical areas may be dictated by soil type. Sandy soils are well suited for range-reared turkeys because they provide good drainage. On sandy soil, ranges that are flat or that have little slope can be constructed. In areas with heavy clay soils, ranging on flat terrain is undesirable because of drainage problems. In such areas, a good ground cover for the open areas, such as fescue or orchard grass, helps both to stabilize the slope and to prevent muddy areas.

Natural cover. The presence or absence of wooded growth or natural cover influences range site selection. Ideally, the range should have both open and shaded areas. Partial shade is extremely important for turkeys reared during the hot summer months. Shade also helps reduce the energy required to lower body temperature. Since 85 to 90 percent of total feed consumed is fed while the turkey is on range, shade becomes an important factor in improving feed conversion and weight-to-age ratios.

Space requirements. The minimum area to range turkeys is governed by the type of soil and degree of drainage. On very sandy soil, up to 1,000 turkeys per acre is acceptable; on clay or clay-loam, 250 to 300 turkeys per acre might be the maximum. These numbers will result in little growth from grazing. For more effective grazing, expect more range area and very good

Natural cover and shaded areas are important for turkeys reared on range.

ground cover per bird. Providing more than the minimum amount of space allows for healthier birds and more efficient management, an ability to move birds and equipment as weather conditions or disease outbreaks dictate, and reduction in the degree of surface contamination. Range sanitation can also be improved by disking or plowing open areas, particularly if pens are to be occupied more than once per year.

Physical Layout of the Range

Physical arrangement is an important aspect of turkey range planning. The layout should be made with two factors in mind: first, the well-being of the turkeys; second, efficiency of management.

Before you put your turkeys on range, do the following:

- Remove all debris, stumps, and limbs from the range area.
- Construct pens in such a manner that turkeys and equipment can be moved up-slope and excluded from previously occupied ground.
- Arrange the area to make sure that surface water does not drain from one occupied pen to another.
- Clear a lane for feeders and waterers if the range is heavily wooded; these lanes should be wide enough to allow efficient movement of the feed delivery vehicle.
- Construct roadways in such a manner that surface drainage from pen to pen is avoided — ideally, a roadway should encircle the range to provide visual access to all parts of the range.

Confinement-Rearing

Rearing birds in confinement has several advantages over range-rearing. It protects against losses from soilborne diseases, predators, thefts, and adverse weather conditions; labor costs and acreage requirements are less; and it reduces the effects of livestock on the environment.

Small-flock producers have numerous housing and management-system options. The brooder house or pen, if large enough, may be used to confine the birds until they reach market age. Several variations from the conventional confinement-rearing systems are used by small-flock owners — the house and porch system is one example. Some growers, after removing the birds from the brooding facilities, confine them to a wire-enclosed porch for added protection. Ideally, a section of the porch should have a solid floor where a dry, fluffy

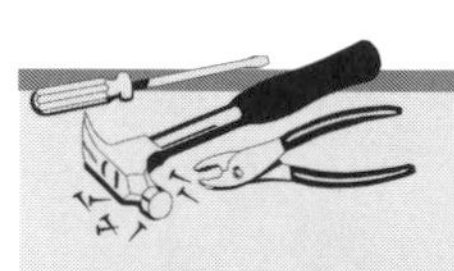

MATERIALS FOR TURKEY SUNPORCH

- 5 pieces of 1 x 4 common board for door, 4' (1.2 m) long
- 8 pieces of 2 x 4 corner posts 5' (1.5 m) long
- 2 pieces of 2 x 4 center studs 5' (1.5 m) long
- 1 piece of 2 x 4 center stud 14' (4.3 m) long
- 5 pieces of 2 x 4 floor joists 8' (2.4 m) long
- 2 pieces of 2 x 4 floor joists 12' (3.7 m) long
- 2 pieces of 2 x 4 plates 12' (3.7 m) long
- 2 pieces of 2 x 4 plates 8' (2.4 m) long
- 2 pieces of 2 x 4 roost 8' (2.4 m) long
- 1 piece of 2 x 4 roost 12' (3.7 m) long
- 1 piece of 2 x 8 rafter 8' (2.4 m) long
- 1 piece of 2 x 4 at feeder 6' (1.8 m) long
- 2 pieces of 2 x 4 cross braces 8' (2.4 m) long
- 33 pieces of 1½ x 1½ fir for flooring, 12' (3.7 m) long (or 96 sq. ft. [8.9 sq m] of 1½ x 1½ turkey wire mesh)
- 26 pieces of 1 x 8 common board, D4S or T & G, 4' (1.2 m) long
- 14 pieces of 1 x 8 common board, D4S or T & G, 5' (1.5 m) long
- 4 pieces of 1 x 12 common board, D4S, 6' (1.8 m) long
- 1 piece of 1 x 12 common board, D4S, 8' (2.4 m) long
- 2 pieces of 1 x 3 common board, D4S, 6' (1.8 m) long
- 1 piece of 1 x 6 common board, D4S, 6' (1.8 m) long
- 1 piece of ½ x ½ strip 6' (1.8 m) long

litter can be maintained. This helps prevent the development of breast blisters or leg and foot problems.

When birds are raised in strict confinement, adequate floor space is important. If the birds' beaks are trimmed, if feed and water space are adequate, and other conditions are optimal, large males can be confined to approximately 5 square feet (0.47 sq m) of floor space, females to 3 square feet (0.28 sq m), and mixed flocks to 4 square feet (0.37 sq m). Smaller varieties need 4 square feet (0.37 sq m) for males, 3 square feet (0.28 sq m) for females, and 3½ square feet (0.33 sq m) for mixed flocks.

50	linear feet (15.2 m) of 2" (5.1 cm) lath for feed rack
1	piece of ¼" (0.6 cm) tempered CDX plywood with exterior glue 3' x 4' (0.9 mx1.2 m)
50	linear feet (15.2 m) heavy galvanized wire, 8 or 9 gauge
3	3½" or 4" (8.9 cm or 10.2 cm) light T-hinges
3	4" (10.2 cm) hasps
3	padlocks
32	linear feet 2"(9.8 m 5.1 cm) poultry wire 48" (122.2 cm) high
12	linear feet 2-inch (3.7 m 5.1 cm) mesh poultry wire 36" (91.4 cm) high
1	roll roofing paper
5	pounds (2.3 kg) 6d nails — com
10	pounds (4.5 kg) 8d nails — com
10	pounds (4.5 kg) 10d nails — com
5	pounds (2.3 kg) 16d nails — com
2	pounds (1.1 kg) ¾" (2.1 cm) galvanized staples
1	pound (0.5 kg) 1" (2.5 cm) galvanized roofing nails

stud = upright
joist = small beams laid horizontally to support floor
plate = horizontal timber carrying rafters for roof
rafter = sloping timber of roof
D4S = dressed/planed on four sides
T & G = tongue and groove
hasp = hinged metal strap secured by staples and pin
6d = 6-penny size
com = common nails with round heads

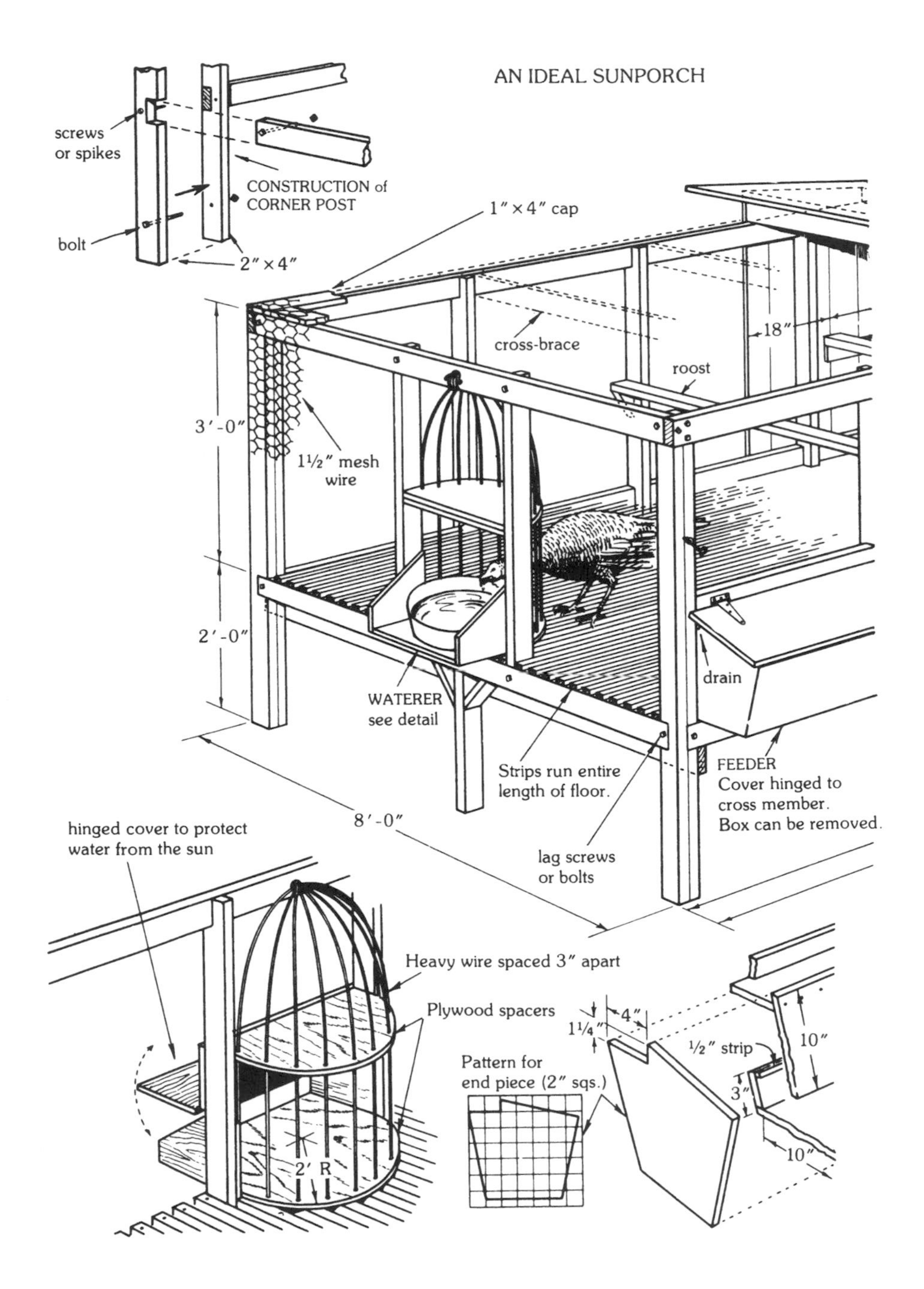

Plan and details for a house and sunporch system suitable for a small flock

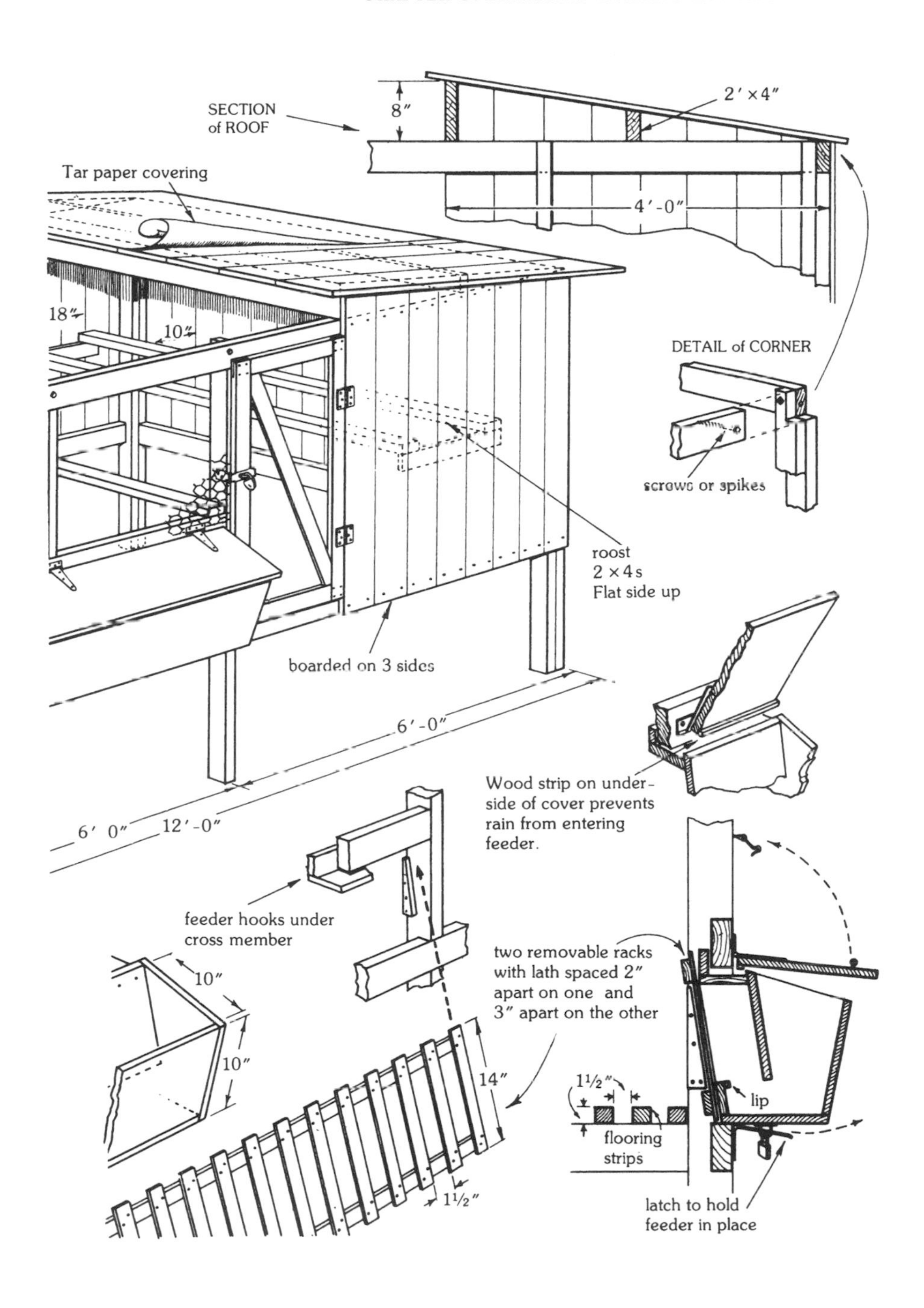
SECTION of ROOF
8″
2′ × 4″
4′-0″
Tar paper covering
18″
10″
DETAIL of CORNER
screws or spikes
roost
2 × 4 s
Flat side up
boarded on 3 sides
6′-0″
6′ 0″
12′-0″
Wood strip on underside of cover prevents rain from entering feeder.
feeder hooks under cross member
10″
10″
two removable racks with lath spaced 2″ apart on one and 3″ apart on the other
14″
1½″
1½″
flooring strips
lip
latch to hold feeder in place

The Yard System

When birds are reared in confinement, they need considerable floor space toward the end of the growing period. By using a yard attached to the housing facility, more birds can be kept in a smaller housing area. The yard should be well drained. You can put gravel or stones in the yard to keep the birds out of the mud, to improve sanitation, and to prevent disease. Yards need to be kept in good condition. Ideally, the location of the yard should be changed every 1 or 2 years. Four to 5 square feet (0.37–0.47 sq m) of yard area per bird is recommended.

If there is a danger of predators, such as foxes or dogs, the yard should be fenced. A woven-wire poultry fence 6 feet (1.8 m) high normally keeps the turkeys inside, but in some cases it may be necessary to clip the flight feathers or primary feathers on one wing (see page 64) to prevent the birds from flying over the fence. Such animals as foxes, raccoons, and dogs can cause considerable damage. The harm is frequently due more to piling and suffocation than to outright killing by the predator. Lighting the range with floodlights also helps keep out unwanted animals and discourages raiding of the range or yard by predators. Electric fencing also serves this purpose.

The Feeding Program

One of the best pieces of advice for turkey growers is to select a good brand of feed and follow the manufacturer's recommendations for its use.

There are two basic feeding programs for turkeys: One is the complete feed system; the other uses a protein supplement plus grains. The latter diet is frequently used in those areas where homegrown grains are available. Feed insoluble matter, such as granite grit, to the birds whenever grains are included as a part of the diet or if the birds are on range. This enables them to grind and utilize the grains and other fibrous materials.

Feed-company recommendations for feeding turkeys vary considerably. However, they are all based on the fact that turkeys grow rapidly. Turkeys need a high-protein diet at the start to support this rapid growth. The nutrient requirements of turkey poults vary with age. As they become older, the protein, vitamin, and mineral requirements decrease and the energy requirements increase.

A Simple Feeding Program

One of the simpler feeding programs for toms begins with a 28 percent protein starter diet fed up to approximately 6 weeks of age. The birds are changed to a 22 to 24 percent protein growing diet and are fed this diet from 6 to 10 weeks of age. From 10 weeks to 14 weeks, a second grower diet that is lower in protein and higher in energy is fed. From 14 weeks until market or slaughter, they are fed a finishing diet containing approximately 16 percent protein. Hens can be fed the same feed, but make changes every 3 weeks after feeding the starter diet through 6 weeks of age. Other feed companies offer and recommend five or six different diets during the growing period.

Feeding Program for Growing Turkeys

			Age of Bird (weeks)	
Feed Type	Form	Protein (%)	Toms	Hens
Starter	mash/crumbles	27–28	0–6	0–6
Grower 1	mash/pellets	22–24	6–10	6–9
Grower 2	mash/pellets	18–20	10–14	9–12
Finisher	mash/pellets	15–17	14–18	12–16

The importance of buying feed from a reputable company and following its recommendations cannot be overemphasized. Some of the early feeds, such as the starter diet, should include a coccidiostat, which is a type of medication that helps prevent coccidiosis (more about this disease in chapter 5). Observe the precautions on the feed tag and the recommended times for withdrawing the feed before dressing birds.

Make sure the birds always have feed and water. When grains are included as a part of the diet, the amount to be fed depends on the protein content of the mash or pellets the birds are receiving. Remember that grains contain low levels of protein (corn has 8 to 9 percent, wheat has 10 to 12 percent, and oats have 11 to 12 percent). Feeding birds too much of these grains dilutes the total protein content of the combined diet to the extent that it could affect growth. When the birds are 12 to 16 weeks old, they can receive a grain and mash diet, but don't dilute the protein content below the 16 percent level. When grains are fed with the finishing diet, avoid diluting the diet with grains to the extent that total protein intake is reduced below 14 percent.

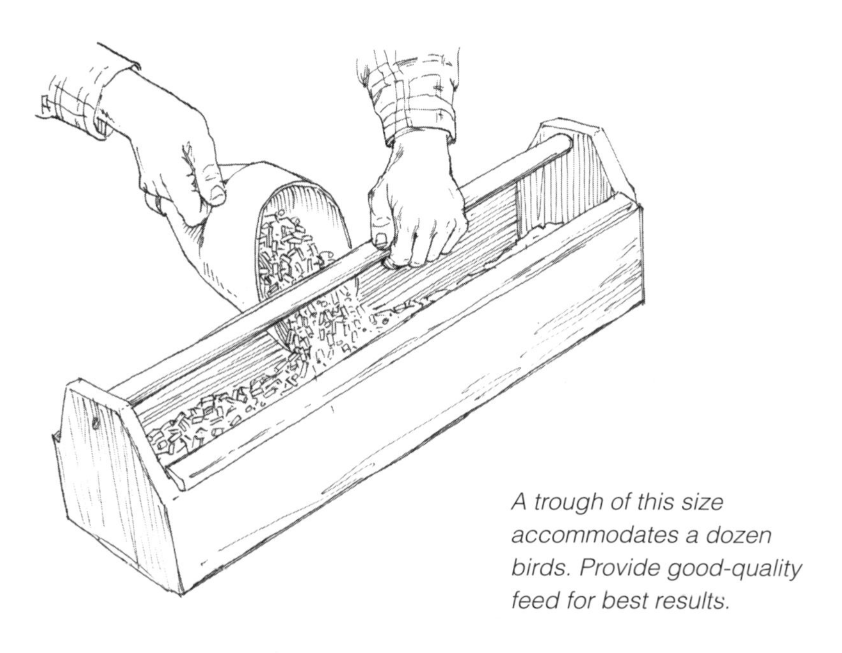

A trough of this size accommodates a dozen birds. Provide good-quality feed for best results.

Check the birds regularly to ensure that they have adequate feed and water.

Most turkey producers feed a nutritionally complete starter mash or crumble. A *crumble* is a pelleted feed that has been reground so it can be eaten by small poults. Poults should not be fed pelleted feed until they are 6 to 8 weeks of age, but you can use green feed for small flocks if labor requirements are not a concern. However, be aware that birds grow best on a complete feed. Any deviation from this may cause a decrease in growth rate, although this decrease may not be a problem for small-flock producers and may reduce feed cost. A tender alfalfa, white Dutch clover, young tender grass, or green grain sprouts, all chopped into short lengths and fed once or twice daily, can be good for poults. Turkeys like tender green feeds, such as short, fresh lawn clippings, and garden vegetables, such as Swiss chard, lettuce, and even the outer leaves of cabbage. Do not let your birds eat wilted, dry, or long, stringy roughage, as this type of feed can cause impacted or pendulous crops. Again, if roughage is fed, make sure the birds receive an insoluble grit, such as turkey-size granite grit.

Percentages of Ingredients for Typical Rations[a]

INGREDIENT	STARTER	GROWER 1	GROWER 2	FINISHER
Corn	44.30	56.00	63.20	66.50
Soybean meal (48%)	40.00	28.00	19.00	15.00
Poultry meal	8.00	7.00	8.00	8.00
Fat	2.50	4.00	6.00	7.00
Dicalcium phosphate	2.40	2.20	1.60	1.50
Calcium carbonate	1.35	1.35	1.10	1.00
Salt	0.35	0.35	0.35	0.35
Choline chloride	0.20	0.10	0.00	0.00
Lysine	0.28	0.40	0.30	0.30
Methionine	0.25	0.18	0.15	0.10
Vitamin premix[b]	0.20	0.20	0.20	0.20
Mineral premix[b]	0.10	0.10	0.10	0.10
Coccidiostat[b]	0.07	0.07		
Total	100	100	100	100
CALCULATED APPROXIMATE ANALYSIS				
Crude protein	28.00	22.00	19.00	17.00
Metabolizable energy (kcal/lb)	1,300.00	1,380.00	1,475.00	1,510.00
Calcium	1.40	1.30	1.10	1.05
Available phosphorus	0.74	0.67	0.56	0.54
Methionine	0.71	0.56	0.50	0.42
Lysine	1.81	1.53	1.20	1.09
Sodium	0.19	0.18	0.18	0.18

[a] The rations given are for the feeding program on page 53.
[b] Use at manufacturer's recommended rate.

Turkeys can be fed the concentrate in mash or pellet form (crumbled for birds less than 6 weeks of age). When you change from a mash to pellets, make the change gradually. A commercial concentrate may also be purchased and combined with ground grain or with soybean meal and ground corn in the proportion recommended by the manufacturer. Usually, small-flock growers find it advantageous to use a complete, ready-mixed mash or crumbled or pelleted feed when the birds are reared in confinement. If the birds are on a good range, a complete feed — preferably in pellet form — or a protein concentrate supplemented with grains and insoluble grit makes a sound feeding program.

Feed Management: The Key to Turkey Health

Feed quality is extremely important to the health of your turkeys. You can optimize performance by following these guidelines:

- Feed nutrients, especially certain vitamins can be destroyed by heat; therefore, feed should be stored in a cool, dry area.
- Use a "batch" or "lot" of feed within 4 weeks of mixing, especially during the summer. Plan the amount of feed to be mixed or purchased based on your rate of use.
- Monitor the quality of your feed ingredients in regard to how they are handled and stored as well as their nutrient content. The final feed product is only as good as the ingredients used to mix the diet.
- Have your feed and feed ingredients analyzed at a laboratory on a regular basis. This is the best way to monitor feed quality.

By the time the effects of poor feed quality are demonstrated in bird performance, it is likely too late for the birds on that feed, and substandard performance will probably be the result. At the very best, you will lose time and profits by having to hold the birds longer than desired to overcome the effects of poor feed. These effects can also be masked by disease or management problems. Contact your local Extension office for more information about feed quality assurance and where to have feed samples analyzed.

Growth Rates and Feed Consumption for Rapid-Growing, Heavy Roaster Turkeys*

AGE	LIVE WEIGHT		TOTAL CUMULATIVE FEED REQUIRED		FEED PER POUND OF LIVE WEIGHT	
(WEEKS)	MALES	FEMALES	MALES	FEMALES	MALES	FEMALES
2	0.6	0.6	0.6	0.6	1.2	1.2
4	2.1	1.8	2.5	2.2	1.3	1.3
6	4.5	3.8	6.4	5.5	1.5	1.5
8	7.9	6.3	12.5	10.4	1.6	1.7
10	12.0	9.1	21.3	17.1	1.8	1.9
12	16.4	12.1	32.6	25.4	2.0	2.1
14	21.1	15.2	46.4	35.4	2.2	2.3
16	26.1	18.3	62.8	44.0	2.4	2.4
18	30.8	21.3	79.9	57.3	2.6	2.7
20	35.3	24.0	100.2	66.5	2.8	2.8
21	37.3			107.8	2.9	
22	39.2		117.2		3.0	
23	40.9		135.3		3.3	

*All amounts given in pounds; 1 lb = 0.454 kg.
Source: Peter Ferket. "Market Rates Up, Growth Performance Steady." *Poultry USA,* Vol. 1, No. 1, January. Mt. Morris, IL: Watt Publishing Co., 2000.

Proper management of the feeders is important. Start out the poults on box tops, plates, or small poult feeders. Provide larger feeders as the birds become older. If this is not done, the turkeys will "beak out" feed or knock over the feeders, and feed will be wasted in the litter. The same is true for waterers. As the birds grow, use larger waterers to avoid spillage and to make sure the poults have an adequate water supply.

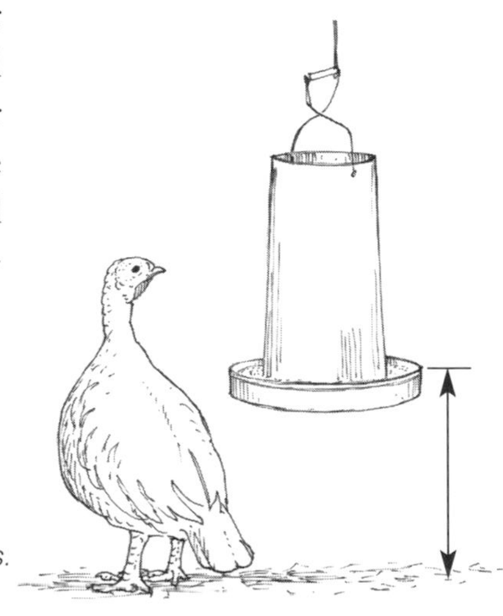

To minimize waste, all feeding equipment should be designed so that it can be adjusted as birds grow. Keep the lip of the hopper at about the level of the birds' backs.

Daily Water Consumption of Large White Roaster Turkeys[a]

AGE IN WEEKS	WATER CONSUMPTION (GALLONS PER 100 BIRDS)
1	1
2	3
3	4
4	6
5	7
6	9
7	11
8	14
9	17
10	18
11	21
12	23
13	26
14–20	27[b]

[a]1 gallon = 3.8 L.

[b]Consumption per 100 birds per day varies, from 15 weeks to maturity, from 20 to 35 gallons (76–133 L) depending on environmental temperatures. A 20°F rise in temperature from 70 to 90°F (21 to 32°C) can cause birds to double their water consumption.

Source: Nicholas Turkey Breeding Farms, Sonoma, CA, 1994.

One of the most common problems seen by servicemen in the field is wasted feed. Although sloppy handing accounts for some of this waste, mismanagement of feeders is the primary reason. Following are several ways to avoid wasting feed:

- Never fill the feed hoppers more than halfway.
- Keep the top, or lip, of the hopper on a level with the birds' backs. This means adjusting feeders or changing to larger feeders as the situation demands.
- Use hanging, tube-type feeders for brooding or confinement-rearing. They may be adjusted easily to prevent waste. One tube feeder that is 16 inches (40.6 cm) in diameter is adequate for 25 birds.

Cannibalism

Feather picking and cannibalism are common problems in turkey flocks, especially when the birds are raised in close confinement.

Factors That Cause Picking and Cannibalism

The following factors contribute to feather picking and cannibalism:

- Overcrowding (including inadequate feed and water space)
- Boredom or idleness
- Temperature too high or too low
- Bright lighting
- Age (more common in young birds)
- Poor sanitation
- Poor ventilation
- Equipment; poorly operating brooders or feeders
- External parasites
- Nutrition — especially amino acid, protein, or sodium deficiency

Overcrowding

Overcrowded conditions not only mean inadequate space for the birds to move around but also involve insufficient feed and water space. Also, make sure birds receive an adequate supply of cool clean water.

Boredom or Idleness

Feather picking or cannibalism may start as a result of boredom. This occurs more often in confined birds than in range-reared birds. Some producers suspend a head of cabbage or lettuce from a rafter or provide a bale of hay for the birds to peck.

Temperature

Brooding temperatures too high or too low might induce the poults to start picking one another. Make sure the brooder temperatures are correct and, more important, observe the birds closely and often to make sure that they are active and seem to be comfortable.

Light

Bright brooder lights tend to increase activity and feather picking, which can lead to cannibalism. Less picking occurs when chicks are brooded under daylight or artificial light of low intensity (½ foot-candle). Use bright lights only at the beginning of the brooding period. Once birds are well started, such as at 1 to 2 weeks of age, decrease light intensity for brooding.

Sanitation and Ventilation

Both poor sanitation and ventilation may cause irritation of the eyes and nostrils, which then become targets for picking by other birds, especially older ones. Keep litter dry and well managed and provide ventilation in the brooder area or in the confinement-rearing area as needed. If air quality is poor for the caretaker, it is probably suboptimum for the turkeys.

Equipment

Poorly designed and arranged feeders and waterers with sharp edges can cause injuries that lead to severe picking and cannibalism problems. Poorly functioning brooders and feeders can lead to picking problems.

External Parasites

Birds that are heavily infested with lice or mites will scratch themselves, and this induces picking by other birds. The parasites themselves can cause severe skin lesions that also induce picking.

Nutrition

Feather picking and cannibalism are often associated with diets that are deficient in certain amino acids, protein, or sodium (salt). Make sure feed is the correct type for the age of the bird; young birds need higher-protein feeds. Do not feed layer mash or scratch feed to turkeys. Sometimes adding fibrous grains, such as whole oats, to the diet helps alleviate picking. Have feed checked for sodium content. The correct level of sodium is 0.15 to 0.18 percent of the diet.

Prevention

Proper flock management helps to prevent these problems. You can also discourage these behaviors by providing adequate shelter, floor space, and feed and water space; eliminating obstacles or equipment that might cause injury; removing dead or sick birds immediately; not introducing new birds into an established population; avoiding frightening the birds; providing a commercially prepared diet and avoiding sudden changes in type or texture of diet; avoiding sudden changes in temperature; and trimming the birds' beaks.

This last practice, beak trimming, deserves special mention because it is probably the most effective means of controlling feather picking and cannibalism. Beak-trimming the poults is best done at the hatchery on the day of hatching and should be a regular practice before noticeable picking occurs.

Trimming is usually done with an electric beak trimmer. The cut can be made with an electrically heated blade that cauterizes as it cuts or with an electrical arc that kills the tissue in the tip of the upper beak. (It may not be economically feasible for small-flock raisers to own an electric beak trimmer. This is one additional reason to purchase poults from a hatchery or mail-order company.) When doing the operation for the first time, it is advisable to get someone to demonstrate the method for you.

When a beak trimmer machine is not available, dog toenail clippers or heavy shears may be used. However, there is some danger of infection and bleeding when the beaks are not properly cauterized. Bleeding can be temporarily controlled by burning the top beak back slightly with a hot iron. When the beaks are trimmed, make sure feed and water levels are kept deep enough so that the birds can consume adequate amounts of both.

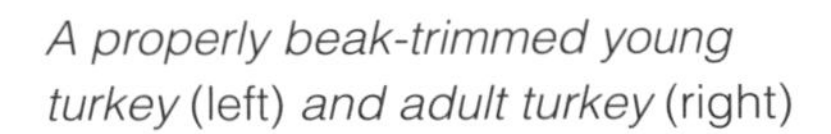

A properly beak-trimmed young turkey (left) *and adult turkey* (right)

For birds going on range, beak trimming is not usually done unless picking and fighting occur. If it is necessary to beak-trim range birds, somewhat less is removed from the beak because excessive trimming interferes with grazing.

Wing Clipping or Notching

Sometimes turkeys tend to fly, particularly when reared in the open or on range. When it is necessary to prevent the birds from flying, the primary wing feathers (flight feathers) of one wing can be cut off with a sharp knife or scissors. Wing notching, or the removal of the end segment of one wing with a beak trimmer, is a drastic and permanent method that is sometimes used to prevent flying. Wing notching can be done with an electric beak trimmer when a poult is 1 to 10 days of age. Older, heavy-type birds, such as the Large White, do not fly.

Toe Clipping

Some turkey producers "toe-clip" their birds to prevent scratches and tears of the skin on the birds' backs and hips that detract from the dressed appearance. Toe clipping is especially helpful in preventing scratches and skin tears when birds are reared in confinement under crowded conditions or if they are nervous. The practice also helps prevent carcass downgrading when birds are on range. The two inside toes are clipped so that the nails are completely removed. Surgical scissors or an electric beak trimmer may be used to remove the toes. Toes should be clipped at the hatchery, but you can do it yourself during the first few days. Small-flock producers may wish to avoid toe clipping. It is an added stress for the poult and, with proper management, is unnecessary.

FOUR

The Breeding Flock

Occasionally, small-flock owners want to keep some of their turkeys as breeders and to produce hatching eggs. The turkey-breeding flock requires time and good management. Not only that, but it is an expensive project as well. The cost of growing breeders to maturity is high (see page 4), and a considerable amount of feed is required to maintain the flock during the holder and breeder periods (see page 66).

Selecting the Breeders

Select breeders from your flock at 16 to 18 weeks of age. Keep the best birds — those that are healthy and vigorous — for breeders. Other important characteristics are good, full breasts (those with nonprotruding keel bones); straight legs, backs, and keel bones; good growth rate; and sound legs and good walking ability.

Breeders are a long-term investment. Therefore, choose individual birds with deliberate care, always keeping the ideal bird and your breeding goals in mind.

Feed Consumption of Turkey Breeders (Pounds per Bird per Day)*

Type of Turkey	Hens	Toms
Large	0.80	1.50
Medium	0.60	1.25
Small	0.50	0.75

*1 lb = 0.454 kg.

Source: *Turkey Production,* Agriculture Handbook No. 393, United States Department of Agriculture.

Mating

You should have about one tom for every ten to fifteen hens. Keep a few spare toms to replace those that die or are poor mating birds. Producers who are serious about hatching-egg production should use artificial insemination. If you have a hobby flock, and if the toms are not too heavy, you can use natural mating. However, egg fertility may suffer somewhat.

Lighting

Turkey hens must be sensitized to light for a period when day length is relatively short, such as during the winter. Breeder hen poults are best reared so that the birds are 16 weeks of age or younger during the fall of the year. This provides the birds with decreasing day length or short days as the birds enter adolescence. The optimum age for photostimulation is approximately 30 weeks for birds of heavy strains during their first year. If females reach this age during the winter or early spring, they will enter into the egg-laying period with good physical maturity. If not, they will lay small eggs for a shorter period. If breeders are kept for several years, they will cycle with the season unless artificial light is provided to increase day length as natural day length decreases.

Day Length

Birds can measure day length. Thus, turkey hens measure the length of time from sunrise, or the time that the lights are turned on, until sunset, or the time that the lights are turned off. If you use natural lighting, calculate day length as being 30 minutes before sunrise to 30 minutes after sunset. Long day length is responsible for both photostimulation of reproduction and photorefractoriness, which includes and is characterized by cessation of reproduction. It is important to note that although photostimulation and photorefractoriness have opposite effects on the hen's reproductive system, they are both natural processes caused by long days. They each have important and beneficial effects on birds in nature (including wild turkeys), but photorefractoriness is a negative occurrence for the domestic turkey breeders because it diminishes overall egg production by shortening the lay period.

Critical day length is the minimum number of hours of light that the hen needs to induce normal egg production. This is generally what is meant by *long day length* — that is, any day length that is longer than critical day length. Although 12 hours of light is generally considered to be a long day length and is stimulatory for turkey hens, it is advised that 14 to 18 hours of light per day be used for optimum stimulation for egg production. Longer day lengths do not necessarily result in greater egg production. If day length is too long, the overall time in production can be reduced by photorefractoriness (diminished response to long day lengths). Therefore, day length should be just longer than critical day length. The goal is to maximize photostimulation and minimize photorefractoriness.

For most parts of the United States, 15.5 hours of light (artificial and natural light combined) is adequate to start hens in the spring. One 50-watt bulb for every 100 square feet (9.3 sq m) of floor space is adequate. You can use a time clock to bracket the natural daylight hours with artificial light. This often prolongs the natural lay period well into the fall. A typically adequate period of egg production could be anywhere from 20 to 30 weeks, depending on the birds and environmental conditions.

The Rest Period

A proper sequence of short-day-length days ends photorefractoriness and restores photosensitivity in turkey hens. In nature, this is accomplished by the short days of winter, and utilizing these short days is probably best for small-flock producers as well. Therefore, hens can be returned to natural day length and allowed to rest during the winter. Hens need approximately 12 weeks for a proper rest period. The reproductive tract regresses, and the birds molt and become reconditioned for a new egg-laying period the next spring.

Light Intensity

The intensity of light refers to the foot-candles of illumination produced by a lightbulb or the sun. Several breeder guides currently recommend 10 to 12 foot-candles as a minimum for breeder hens. Because of the effectiveness of natural light in stimulating egg production, there is probably no upper limit for artificial light intensity.

Altering Seasons and Shortening Days

The seasons of the year can be altered for turkeys by using light-tight buildings with good ventilation and artificial light. In addition, light traps must be used in conjunction with power ventilation (fans) if you want short day lengths for the birds when the natural day length is long. However, this is usually not the goal for most small-flock producers.

Options for Lightbulbs

Numerous types of lightbulbs are available for poultry: among them, high-pressure sodium, fluorescent, and incandescent. Turkeys respond to all three types. The high-pressure sodium and fluorescent bulbs are more efficient than incandescent bulbs. The small-flock producer may find it difficult to locate high-pressure sodium or even fluorescent bulbs designed for poultry, but poultry-supply companies now provide a compact fluorescent lightbulb. These bulbs come with a ballast and can be screwed directly into a regular socket. Although these bulbs have a higher initial cost, studies show that they are less costly in the long run because they use one-fifth the energy required by incandescent bulbs and last longer.

Photostimulation

Approximately 8 weeks prior to mating, stimulate the toms with at least 14 hours of light per day. Do not let day length shorten for the toms or hens. Light stimulation is needed to get good sperm production from the toms and good egg production from the hens. If day length for when the toms will be bred is longer than 14 hours, add light as necessary in the confinement area to match natural day length. If natural day length begins to shorten, add enough artificial light time to maintain a 14-hour day. Hens should be photostimulated with long days 3 weeks prior to the desired onset of egg production. Toms respond to light more slowly than do hens, thus their lighting period needs to be initiated before that of hens. It has been suggested that day length for toms not exceed 18 hours, as excessive lighting periods can cause molting and decreased semen production.

The simplest plan for small-flock producers is to let the birds rest naturally during winter, come into production during spring with increasing natural day length, and maintain long days into late summer/early fall with artificial light.

Egg Production

Most varieties of turkeys can be expected to lay 80 to 100 eggs per bird, especially when artificial light is used to prolong day length. If only natural day length is used, then egg production will be greatly reduced. Egg production is best the first year and diminishes by 20 percent for each succeeding year. Egg size increases the second year, but hatchability tends to decrease.

Mating Habits

Light-type turkeys can mate naturally. Hens begin to mate when egg production commences. If the toms have been exposed to increased day lengths, then they will be able to effectively fertilize the hens. During the breeding period, strutting and courting activity increases among the males. Then hens select the toms of their choice and squat near them. The male mounts the hen, and copulation usually occurs. The tom may mount the female

Breeder tom demonstrating strutting and courting behavior

but not complete insemination. In these cases, the hens may lose interest and not mate for some time; this obviously reduces the fertility rate. Continual mating can be rough on the hens; a tom can injure the hen's back, in some cases severely.

Two breeder hens display squatting behavior, a clear sign that they are ready for breeding.

Artificial Insemination

If you're serious about producing hatching eggs, artificial insemination is the way to go, especially for heavy turkeys. Heavy toms cannot mate naturally. Their breast is too large, and they do not have the proper balance or agility. If your toms are of this type or body size, then you have to use artificial insemination. All commercial turkeys are artificially inseminated. Even if the turkey can mate naturally, artificial insemination increases fertility and therefore hatchability. Hens are sometimes inseminated artificially to supplement fertility if natural mating has not produced good results. Artificial insemination also reduces wear and tear on breeder hens because they are not continuously mounted by the toms. The toms and hens are kept separate; semen is collected from the toms and then administered to the hens.

Collecting Semen

Semen may be collected from the toms two times per week. To work, or "milk," a tom requires two people.

1. One person places the bird on a padded table or a bench with the breast resting on the surface.
2. The primary person, or "milker," as he is known in the turkey industry, stands by the table or sits on the bench. The tom is in front with its head to the left and its tail to the right for a right-handed person. The helper holds the legs together and downward.
3. The milker then places the left hand on the back of the tail and the right hand on the ventral part of the tail or rear part of the abdomen.
4. The milker stimulates the tom by stroking the abdomen and pushing the tail upward and toward the bird's head with the right hand.
5. The male responds, and the copulatory organ enlarges and partially protrudes from the vent. If the copulatory organ does not protrude, the tom is probably not sexually responsive yet and needs more time. ▶

6. If the bird does respond, the milker brings the left hand under the right hand and pinches off the cloaca at the walls of the vent with the left forefingers and thumb while the copulatory organ is exposed. The right hand is used to provide inward and upward pressure beneath the cloaca.

7. The semen is then squeezed out with a short, sliding, downward movement of the left hand and an upward pressure of the right hand. Imagining a scooping action, especially with the left hand, might aid the process. ▶

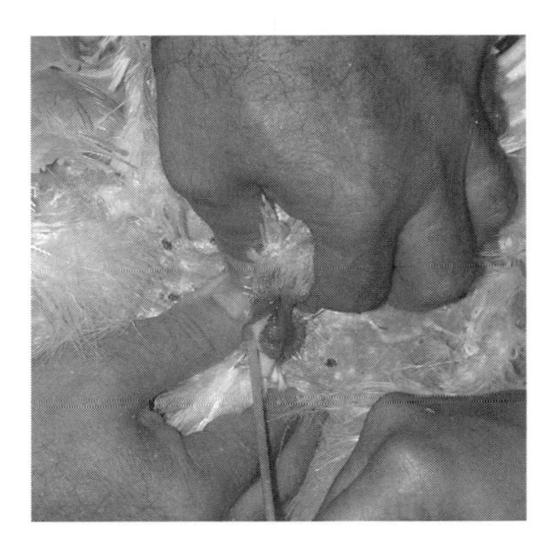

Do not touch the copulatory organ during collection. Toms become trained quite quickly and ejaculate easily when stimulated. A tom produces 0.2 to 0.5 mL of semen per collection.

The semen should be milky in appearance and must be free of fecal matter. Some contamination can be avoided by withholding feed from the toms 8 to 12 hours before semen collection. Also, practice improves a handler's success rate in both handling the toms and collecting clean semen.

Semen is collected in a small syringe (without the needle), such as a 1 mL tuberculin syringe or a small, clean, dry testtube, glass beaker, or a stoppered funnel. Unlike some types, turkey semen cannot be held long and ideally should be used within 30 minutes of collection. Hens can be successfully inseminated with 0.05 mL of semen. However, inseminating an amount this small is difficult for small-flock producers to do without special equipment, which is expensive. Using an inexpensive 1 mL tuberculin syringe, you can administer 0.1 mL of semen, twice the recommended amount, quite accurately. Good results are achieved when hens are inseminated twice within 10 days of egg production. Insemination should be done weekly thereafter for optimal fertility or least every 2 weeks for hobby flocks.

Insemination

Again, insemination is best done with two people. The objective is for one person to evert the oviduct while the other person inseminates the hen. There are several ways to handle the hen to evert the oviduct, and each handler can experiment to find a suitable way. Two possibilities follow.

Option 1

1. Pick up the hen with both legs in the left hand if you are right-handed.
2. You may wish to lean back against a support. The hen's breast may or may not rest up against your knee, left or right.
3. Place your right hand on the tail of the hen so that your palm is to the right of the vent and your fingers are above it and your thumb below it.
4. With your right hand, press the tail of the hen toward its head to evert the oviduct. Rotate your hand away from your body. Using downward pressure with your thumb helps the process.
5. The oviduct will come into view on the left side of the cloaca; this is both the bird's and the handler's left.

Option 2

If you prefer, you can evert the oviduct in this way:

1. Pick up the hen with both legs in the left hand if you are right-handed.
2. Sit with the bird facing you, with the hen's breast resting on your lap.
3. Expose the oviduct by exerting pressure on the abdomen while simultaneously forcing the tail upward toward the head. The oviduct can be exposed only in hens that are in laying condition.

Once the oviduct is everted, the inseminator then inserts a small syringe without the needle 1½ inches (3.8 cm) into the oviduct. As the handler releases the pressure on the vent, the inseminator forces the semen into the oviduct and removes the syringe or tube. Special devices, such as glass tubes or plastic straws, may also be used to inseminate. In this case, a rubber tube is attached to the straw. The straw is placed in the oviduct and the inseminator blows the semen into the oviduct as the pressure on the vent is released. Sperm can be stored in the oviduct for several weeks. However, fertility is at its peak during the first days following insemination.

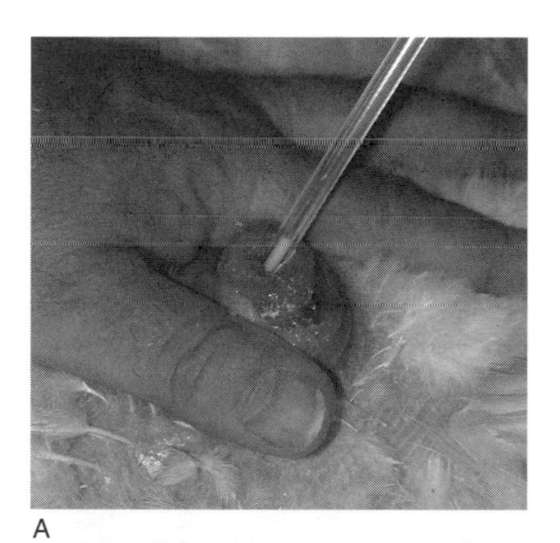

A

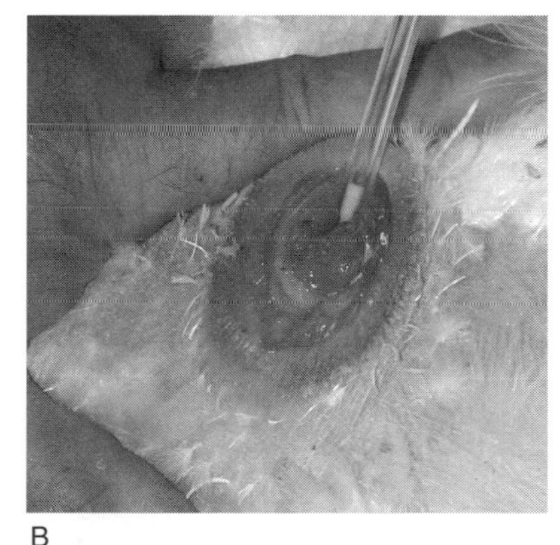

B

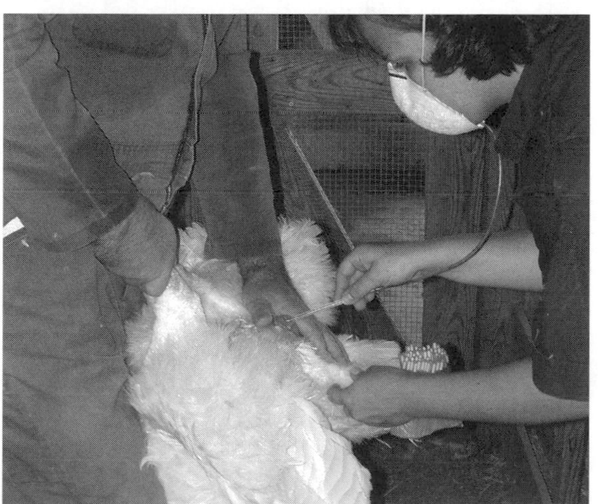

C

A. *The oviduct is everted and the syringe is inserted.*

B. *The handler releases the pressure on the vent and the inseminator forces semen into the oviduct.*

C. *A handler* (left) *and inseminator* (right) *work together to inseminate a hen.*

Fertilization

After successful insemination, the sperm swim up the oviduct to storage sites located in the infundibulum and the junction of the shell gland and the vagina. Fertilization takes place in the infundibulum as the ova (yolks) pass through.The egg continues down the oviduct, passing through the magnum, isthmus, shell gland, and vagina. It then travels through the cloaca and out the vent during egg laying.

Egg Formation

Most of the albumen is contributed to the egg by the magnum. The inner and outer shell membranes are added in the isthmus. The shell gland is distinct from the rest of the oviduct because it is thick, muscular, and darker in appearance. (The shell gland is often referred to as the uterus. However, the shell gland of a bird operates differently and has functions unlike those of the uterus of a mammal.) More fluid is added in the shell gland, but the major function is deposition of the shell and then the cuticle. It takes 24 to 26 hours for egg formation to occur from the time the yolk is released from the ovary until it is laid, and most of this time — at least 20 hours — is spent in the shell gland.

Broodiness

Broodiness is the tendency of the hen to set on, or want to hatch, eggs. When broodiness is permitted to go unchecked, egg production suffers. Turkeys are genetically inclined toward broodiness. Nests should be checked in the early morning or evening for broody hens. These birds should be removed from the nest immediately and broken of the habit by removing the nests. Special pens with slat or wire floors are best for broody birds. Broodiness may also be discouraged in several ways.

- Move the birds to different areas.
- Provide roosts.
- Chase birds off the nest when eggs are gathered; this prevents hens from roosting in their nests.
- Collect eggs several times a day.
- For confined hens, provide enough light (10–12 foot-candles) and make sure there are no dark areas or corners; hens tend to sit in these areas, and this increases the likelihood that they will become broody.

Housing and Breeders

Buildings used for growing poults can also be used for breeding birds. Although adult turkeys don't necessarily need a warm house, it's best if the house is well insulated and ventilated for maximum comfort during cold or hot weather. Floors can be concrete, wood, or dirt. Clean and disinfect floors thoroughly between flocks. Place a good covering of litter on the floor. Usually, roosts or porches are not used for breeding flocks. The house needs electrical wiring because stimulatory light for the breeders is required when days are short.

Nests

Make nests available before egg production begins to give the hens an opportunity to become accustomed to them. One nest for every two to four birds is sufficient. Locate the nests in areas with subdued light, but not dark areas. Dark areas, such as the corners of the pen, attract the birds and encourage broodiness.

Open-type nests are satisfactory for small flocks. Trap nests or tie-up nests are available for producers who want to do some controlled breeding work. Semitrap nests are also available. These nests allow one hen at a time into the nest. When the hen enters, the trap closes so other hens cannot enter. When it leaves, the trap is left open so that other hens can have access to the nest. Most small-flock or hobby-flock owners do not need trap or semitrap nests.

When building a nest, keep the following in mind:

- Nests that are 24 inches (61.6 cm) high, 18 inches (45.7 cm) wide, and 24 inches (61.6 cm) deep provide plenty of room for even large-type hens.
- A foot-board approximately 3 to 5 inches (7.6–12.7 cm) high on the front holds the nesting material.
- Barrels or boxes can be used.
- If for any reason the nests are outside the shelter, cover them to protect them from the elements.
- Nests can be placed directly on the ground as long as they are not in a low spot.
- Nests can be off the ground as long as ramps are provided so that all hens, including especially large ones, can walk up the ramps and into the nests.
- Good nesting materials are wood shavings, chopped straw, sugarcane, and rice hulls.

Simple nests can be effective. Notice that these nests are slotted, which allows entrance of light and air.

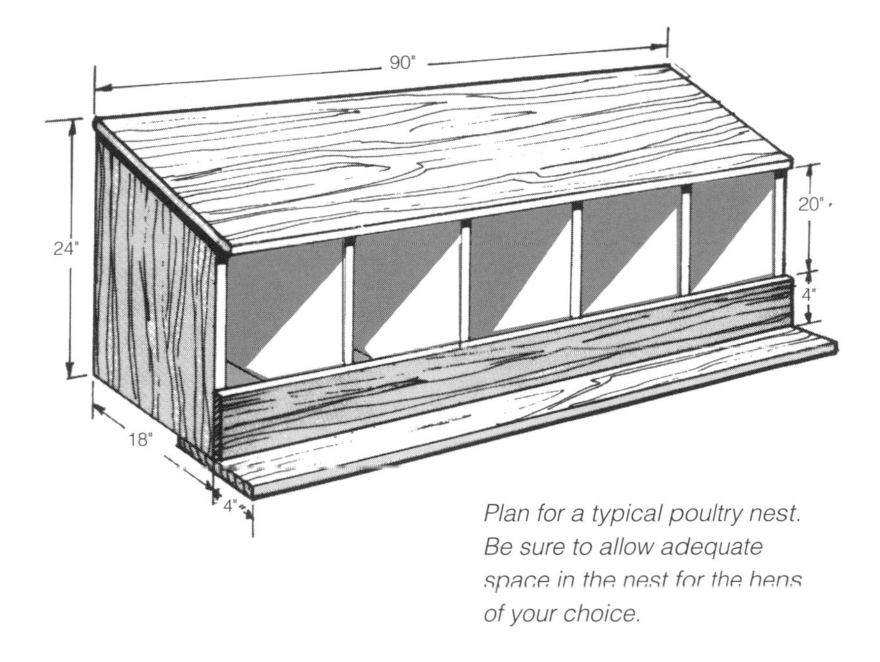

Plan for a typical poultry nest. Be sure to allow adequate space in the nest for the hens of your choice.

Space Considerations for Breeders

Breeders require more floor space than growing birds do. Six to 8 square feet (0.56–0.74 sq m) per bird is the usual recommendation. When hens are housed separately from the toms, 5 to 6 square feet (0.47–0.56 sq m) of floor space is adequate for large turkeys and 4 to 5 (0.37–0.47 sq m) is good for small ones. For areas in which the climate is warm or the winters are mild, breeders may be allowed access to a fenced range or yard area. On range, provide 150 square feet (13.95 sq m) of area per bird, a yard should provide 4 to 5 square feet (0.37–0.47 sq m) per bird, and the shelter should provide at least 4 square feet (0.37 sq m) of space per bird. Locate feeders, waterers, and the broody pen in the shelter.

Feeding the Breeders

Two weeks before the onset of egg production, put the breeders on a turkey breeder diet. Follow the feed manufacturer's instructions. The turkey breeder diet is usually formulated to provide 16 to 18 percent protein and 2.5 to 3 percent calcium. Complete pelleted diets are preferred, but protein concentrate and grain diets may be used if fed in the right proportions. It is imperative that the hens receive adequate calcium for proper eggshell formation. If too much grain is fed in relation to the complete feed or concentrate, egg production and hatchability will be reduced. Hens can be provided with free-choice oyster shell to supplement calcium in the feed. The protein supplement and grain system has the advantage of making use of homegrown grains without the need for grinding or mixing. Homegrown grains can also be used in a complete feed program by grinding and mixing with a concentrate at home or by having them custom-mixed elsewhere. During the winter or any other rest period, the hens can be fed a holder (maintenance) ration. This ration has less protein and calcium than a breeder ration.

Feeding Breeder Toms

Breeder toms of heavy strains can become quite large and difficult to handle. Therefore, it may be useful to restrict the amount of feed these birds receive. This is easy if the toms are kept separate from the hens. Once the toms are mature and are producing semen, feed them on a daily basis to maintain their weight or to allow them to gain weight slowly by using the restricted-fed ration given on page 81. Alternatively, once the toms are producing semen, they can be full-fed a pelleted diet (see page 81) made from corn with vitamins and minerals. This provides plenty of energy and protein for maintenance but prevents rapid growth.

Large turkey breeders need 6 inches of feed hopper space or the equivalent. Smaller birds can get by with 4½ inches (11.4 cm). Follow the manufacturer's recommendations for tube or other types of feeding equipment.

Percentages of Ingredients for Typical Breeder Turkey Rations

	Hen	Hen	Breeder Toms	
Ingredients (%)	Breeder	Holder	Full-fed	Restricted-fed
Corn	64.1	77.80	92.00	73.00
Soybean meal (48%)	22.5	11.50	—	14.50
Fat	4.0	2.00	2.00	2.00
Alfalfa meal or wheat midds	—	5.00	3.00	5.00
Calcium carbonate	6.5	1.00	—	2.75
Dicalcium phosphate	2.2	1.80	2.00	2.40
Methionine	0.1	0.25	0.25	—
Salt	0.3	0.30	0.30	0.30
Vitamin premix	0.2	0.20	0.20	0.20
Mineral premix	0.1	0.10	0.10	0.10
Total	100	100	100	100
Calculated Approximate Analysis				
Crude protein	16.0	12.5	8.00	14.0
Metabolizable energy (kcal/lb)	1,350.0	1,420.0	1,480.00	1,330.0
Calcium	3.0	0.9	0.47	1.6
Available phosphorus	0.5	0.4	0.46	0.6

Hatching the Eggs

Gather the eggs *at least* three times daily. This should be done even more frequently if the birds tend to use certain nests more than others and during extreme weather conditions, such as

Store eggs in egg flats to prevent breakage. One egg is set on top here to show its coloratiion and shape. Always store eggs large-end up.

hot, humid summer days. Frequent gathering helps avoid breakage, excessively dirty eggs, possibly frozen eggs, and broodiness. Discard eggs that are badly soiled, cracked, or soft shelled. Slightly or moderately dirty eggs can be washed in a detergent sanitizer that is formulated specifically for washing eggs. Wash at a temperature of 110° to 115°F (43.3°–46°C) for 3 minutes. Excessively dirty eggs gathered in warm, humid weather may easily become contaminated. Contaminated eggs frequently explode in the incubator, and this contaminates all of the other eggs or poults.

Wash dirty eggs only. Clean eggs that are in clean nests and are stored properly probably do not need to be washed. Washing clean eggs with dirty eggs contaminates the clean eggs. If you have only a few dirty eggs, you may consider just not using them. Washing gets rid of the obvious contamination; however, bacteria may have already penetrated the shell and cannot be removed by washing. Dirty eggs that are cleaned and then incubated may still go bad, rot, and then explode. Your efforts may be best spent encouraging hens to use clean nests rather than on washing and cleaning dirty eggs.

Washing Eggs

Washing can be done with a basket washer that agitates the water or rotates during the washing process. Don't wash the eggs longer than 3 minutes. After washing, rinse the eggs in water that is the same temperature or slightly warmer than the wash water. The water should contain an approved sanitizer, such as one of the quaternary ammonium compounds, at a concentration of 200 parts per million (ppm).

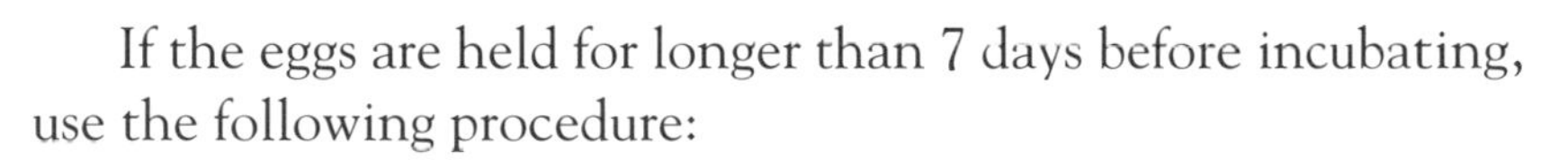

If the eggs are held for longer than 7 days before incubating, use the following procedure:

1. Place the eggs large-end up in a carton or flat.
2. Prop one end of the carton at an angle of approximately 30 degrees.
3. Shift the carton so that a different end is propped every 24 hours; hatchability improves if this is done.
4. Keep eggs that are being held for more than a day in a storage area at a temperature of 55° to 70°F (12.8°–21°C) and 75 percent relative humidity.
5. Store the eggs with the large ends up. Storage under proper conditions for 2 to 3 days may actually improve hatchability.

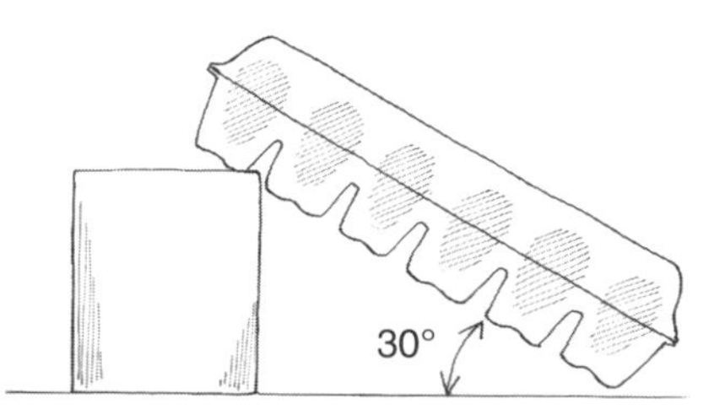

Storing eggs large-end up in a carton that is propped up 30° may improve hatchablility.

Hatching with Broody Birds

Eggs can be hatched by setting them under broody birds. Chickens, turkeys, and even ducks or geese can be used. It is more efficient to keep the hens laying eggs and to hatch the eggs by some other means. A medium-size broody chicken can cover six or seven turkey eggs. Select a calm bird, one that is not likely to be easily frightened or to break the eggs.

Provide a suitable nest box — one that is roomy and deep enough that the broody hen has ample room to turn the eggs, change position, and be comfortable during the 28-day incubation period. Select an area where she will be by herself or with other broody hens. She should have food and water available at all times.

It may be advisable to put the broody hen on dummy or artificial eggs for a few days to make sure she's a persistent brooder. Try taking her off the eggs a few times. If she immediately goes back on the eggs, put the real ones under her. Check broody birds for lice and mites before setting, and treat them for these pests if necessary.

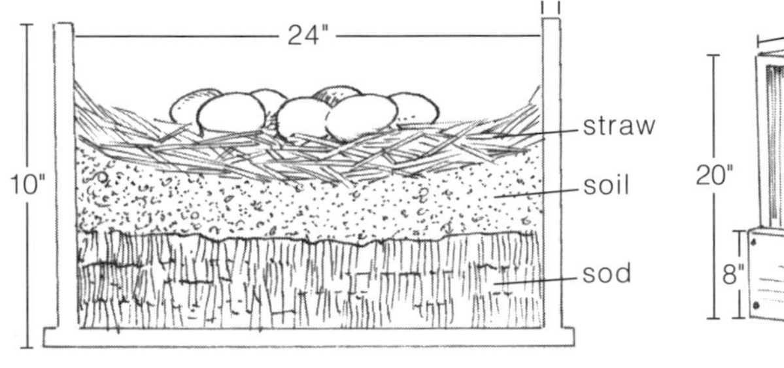

Construction of a nest

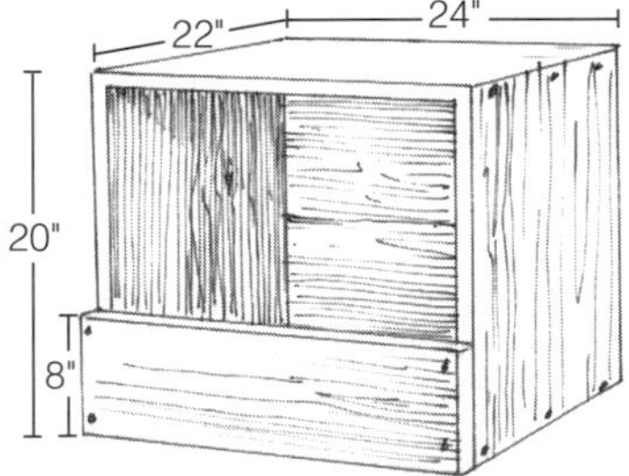

Nest box for a broody hen

Hatching with Artificial Incubators

Two types of incubators are used today: the forced-draft machine and the still-air machine. The large commercial hatcheries use forced-draft machines. The capacity varies from a hundred to several thousand eggs; some of the earlier mammoth machines held approximately 75,000 eggs. This type of incubator has fans that force air through the machine and around the eggs. For most types of eggs, the temperature setting of the forced-draft machine is 99.5° to 99.75°F (37.5°–37.6°C).

Most still-air machines are quite small. The egg capacity ranges from one to about one hundred eggs. Still-air machines do not have fans but depend instead on gravity to circulate air through vents on the top and bottom of the machine. The operating temperature of the still-air machine is higher than that of the forced-draft. It ranges from 101.5° to 102.5°F (38.6°–39.2°C), depending on the type of egg being set. Reasonably good results can be obtained by using an operating temperature of 102°F (38.9°C). The temperature may vary between 100° and 103°F (37.8° and 39.4°C) but shouldn't stay at either extreme. The temperature in still-air incubators should be measured at the top of or above the eggs.

Forced-draft incubator. The eggs are placed in wire bottom trays or special racks designed to permit air circulation around the eggs. They stay in these trays until 3 or 4 days before expected hatch and then are placed in special hatching trays for the remainder of the incubation period.

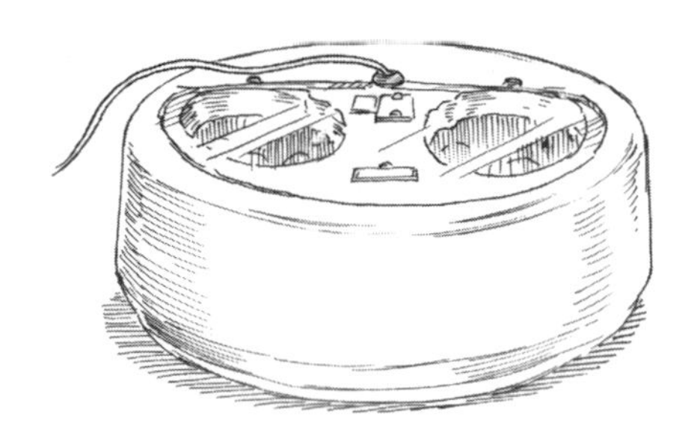

Still-air incubator

You can make an incubator at home. However, several poultry-supply companies provide good incubators for the small-flock producer.

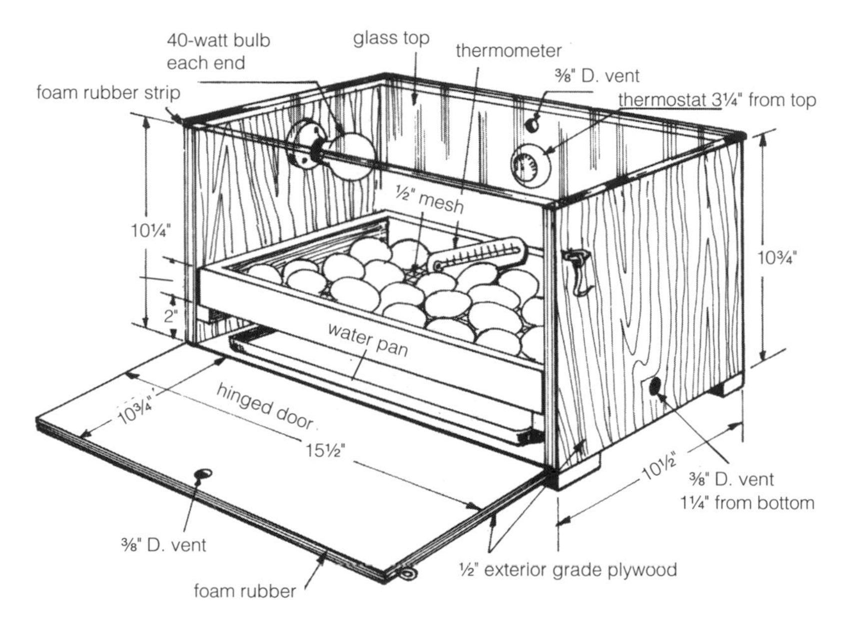

Plan for a simple homemade incubator (D. = diameter.)

The Incubation Period

The normal incubation period for turkeys is 28 days. The first 24 days are frequently referred to as the *incubation period*, and the last 4 days are known as the *hatching period*.

During the incubation period, the eggs lose weight through evaporation at a rather consistent rate depending on the humidity in the incubator. Eggs may be bulk-weighed at the time of setting and periodically thereafter as a check on humidity conditions; this can be used as a rough guide to control humidity in the absence of a wet-bulb thermometer. Humidity in the incubator can be adjusted up or down to compensate for too much or too little weight loss.

Humidity and the Incubator

In a forced-air machine, a wet-bulb thermometer is used to measure humidity. The relationship between wet-bulb temperature and dry-bulb temperature is commonly described as *relative humidity*.

Most incubators are equipped with both a dry-bulb and a wet-bulb thermometer. Relative humidity tables have been established that indicate the relative humidity when the temperature difference between the dry-bulb and the wet-bulb thermometers is at various levels.

For the first 25 days of incubation of large turkey eggs, relative humidity should be approximately 55 percent. At a dry-bulb reading of 99.5°F (37.5°C), the wet-bulb reading should be 85°F (29.4°C). After the eggs are transferred to the hatching compartment, at around 25 days, the relative humidity should be about 70 percent. If the hatcher's temperature is 98.5°F (36.9°C), the wet-bulb reading should be about 90°F (32.2°C). The humidity in small machines is usually furnished by evaporation pans. Humidity has to be controlled by the amount of area of the evaporation pans or by controlling the ventilation. When adding water to the evaporation pans, use warm water to avoid reducing the incubator temperature. If the humidity is too high, open ventilators slightly; if humidity is too low, cut down on the amount of ventilation.

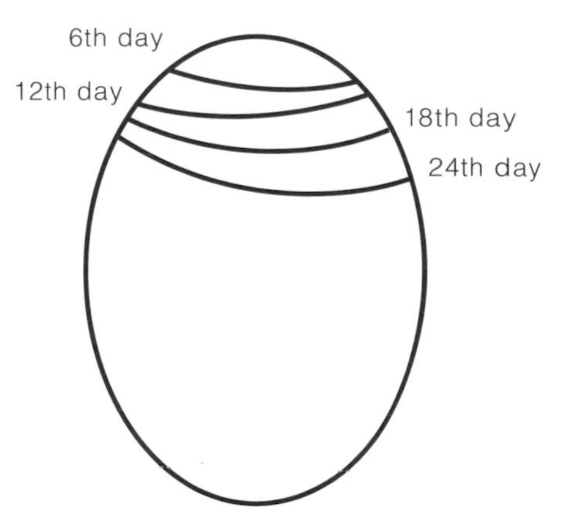

Approximate size of air cells at various stages of incubation when evaporation (weight loss) is normal

Weight Loss during Incubation

DAYS OF INCUBATION	WEIGHT LOSS (%)
6	2.5
12	5.0
18	7.5
24	10.0

Source: *Turkey Production,* Agriculture Handbook 393, United States Department of Agriculture.

Turning the Eggs

During incubation, the eggs should be turned 3 or 5 times daily — always an odd number. The purpose of turning is to prevent the embryo from sticking to the shell membrane. Some machines are equipped with automatic turning devices. Turkey eggs do not have to be turned after 21 days of incubation.

If eggs are packed in the trays large-end up, the trays should be slanted about 30 degrees. If the eggs lie flat on their sides, they should be turned 180 degrees. One method of making sure all eggs are turned is to mark them with an X on one side of the egg and an O on the other, in pencil. When turning, keep all the Xs or Os on top. Discontinue turning the eggs the last 3 or 4 days of the incubation period.

Candling

Frequently, with small machines the eggs are candled 3 or 4 days prior to hatching and the infertile eggs are removed. Candling is done in a dark room by using a special light. If the eggs are infertile, they appear clear before the light. Fertile eggs that have incubated for 24 to 25 days permit light through only the *air cell*, or large end of the egg. The remaining part of the egg appears black or very dark in color. Eggs may be candled at 72 hours to determine fertility. At that point, fertile eggs have a typical blood vessel formation that looks like a spider.

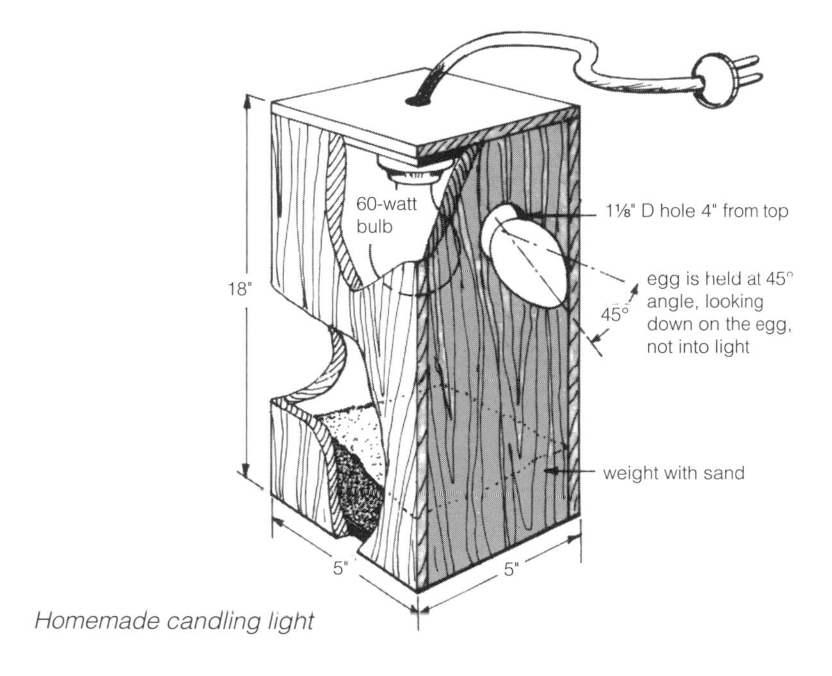

Homemade candling light

Hatching

When the eggs hatch, leave the poults in the incubator for approximately 12 hours, until they are dried and fluffy, before removing them. They can survive for approximately 3 days without food or water, but the sooner they are put on feed and water, the better.

During the latter stages of incubation, the remaining yolk (which is in a yolk sac) is internalized by the poult. The hatched poult can use this yolk for nourishment, particularly for growth processes. However, the poult will be stronger and healthier if food and water is provided as soon as possible.

Poults should not be held too long in the hatcher to avoid excessive drying; dehydration can stress the newly hatched poults. If the poults are to be held for some time before being put in the brooding facility, they should be put into poult boxes and placed in a 75° to 80°F (23.9°–26.7°C) holding room to prevent chilling; the relative humidity of the room should be approximately 75 percent to avoid dehydration.

Incubation Troubleshooting Chart

PROBLEM	PROBABLE CAUSE	SUGGESTED REMEDIES
Eggs clear — no blood ring or embryo growth	Males sterile	Careful culling; select for high hatchability
	Males too old	Do not use
	Males not mating successfully	Use artificial insemination
	Eggs too old, chilled, overheated, or held too long	Set within 10 days, collect frequently, store at 50° to 55°F (10.9° – 12.8°C), 75% relative humidity
	Birds too closely confined	Provide adequate floor space
	Seasonal decline in fertility	Use early-hatched males timed for best maturity or artificial insemination
	Inadequate nutrition or water	Feed hens a fresh breeder diet; provide adequate waterers, well distributed; add vitamins to water
Eggs appear clear when candled, but show blood or very small embryo when broken out	Incubator temperature too high or too low	Check thermometer; operate at correct temperature
	Badly chilled or overheated or held too long	Collect eggs frequently; store at 50°F (10.9°C), 75% relative humidity
	Breeding flocks out of condition	Do not set eggs from sick birds, birds in poor condition, or those that have been recently vaccinated
	Hereditary poor hatch ability	Select strain known to have high hatchability
	Improper nutrition	Feed breeder ration of good quality
Many dead germinating chicks	Faulty incubator temperature	Check accuracy of thermometer; operate at correct temperature
	Lack of ventilation	Provide plenty of fresh air in incubator room and good ventilation in the incubator
	Improper turning	Turn eggs four times daily
	Low-vitamin rations	Feed high-quality breeder rations
Poults fully formed but dead without *pipping;* may have considerable quantities of unabsorbed yolk	Low average humidity in incubator; humidity too low or too high at transfer time in hatcher	Maintain proper humidity throughout incubation and hatching cycle

PROBLEM	PROBABLE CAUSE	SUGGESTED REMEDIES
Poults fully formed but dead without *pipping;* may have considerable quantities of unabsorbed yolk (continued)	Temperature, ventilation or improper turning	Follow recommendations on temperature and ventilate room and machine. Turn four times daily.
	Chilled eggs	Gather eggs frequently; hold under proper conditions
	Disease or flock in poor condition	Diagnose disease and correct flock problems
Eggs pipped but poults dead in shell	Low average humidity	Ensure that wet-bulb temperature is 85° to 90°F (29.4° to 32.2°C)
	High temperature for short time	Maintain recommended temperatures throughout hatch
	Poor ventilation	Provide adequate ventilation of the incubator room and proper openings of the incubator and hatcher ventilators
	Low average temperature	Maintain the recommended temperatures throughout hatch
Poults sticky or smeared with egg contents	Low average temperature at hatch	Use proper temperature
	Average humidity too high	Maintain proper humidity levels
	Inadequate ventilation	Ensure adequate incubator room ventilation and humidity
Shell sticking to poults	Eggs dried down too much	Ensure proper ventilation and humidity
	Low humidity at hatching	Ensure proper humidity wet-bulb 85°F (29.4°C) until pipping — then increase to 88° to 90°F (31.1° – 32.2°C)
Poults hatching too early; navels bloody	Temperature too high	Maintain proper temperature levels throughout incubation to hatching
Rough navels	High temperatures or wide temperature variations	Maintain proper incubator temperature throughout incubation and hatching
	Excessive humidity in hatcher	Use less humidity first 24 to 36 hours after transfer
Poults too small	Low humidity	Maintain proper humidity levels
	High temperature	Maintain proper temperatures
	Small eggs	Don't set those that are too small

Incubation Troubleshooting Chart (continued)

PROBLEM	PROBABLE CAUSE	SUGGESTED REMEDIES
Large, soft-bodied poults	Low average temperature	Maintain proper temperature
Weak poults	Too hot in the hatcher	Watch temperature in hatcher, especially after hatch is complete
	Inadequate ventilation in hatcher	Check ventilators; open wider as hatch progresses
	Breeder flock condition	Diagnose and correct flock problems
Short down on poults	High temperature	Maintain proper temperature throughout
	Low humidity	Maintain proper humidity
Sluggish hatching; some poults begin early but finish hatching slowly	Improper handling of hatching eggs	Gather frequently; store properly
	Improper temperature	Maintain proper temperature throughout
Crippled and malformed poults	Cross beak heredity	Careful flock culling
	Missing eye — rare but may be due to high temperatures	Matter of chance
	Wryneck — nutrition suspected	Check breeder nutrition, especially vitamins
	Crooked toes	Maintain proper temperature
	Sprawled legs — smooth surfaces in hatching trays	Use crinoline cloth in hatch trays
Malformed poults; excessive malpositions	Eggs chilled	Gather frequently; hold under proper temperature and humidity conditions
	Improper turning	Turn 3 or 5 times daily
	Inadequate ventilation	Provide adequate room ventilation; set ventilator controls on machine
	High or low incubator temperatures	Maintain correct temperatures
	Low humidity	Maintain proper humidity level
	Improper nutrition	Use high-quality breeder diets

Source: Based on material provided by the Robbins Incubator Co., Denver, CO.

FIVE

FLOCK HEALTH

A good biosecurity program is important for rearing turkeys or any other livestock. *Biosecurity* is an attitude, program, or management process that provides your birds with a rearing environment that is safe from all hazards and especially those related to disease. Biosecurity is important for all poultry flocks regardless of flock size.

Disease Basics

Disease is a departure from a healthy state and includes any condition that impairs normal body functions. Infectious agents, including bacteria, viruses, fungi, and parasites, that cause disease in poultry can be introduced into a flock. Normally, disease losses are rare in small turkey flocks. However, there are several diseases that may possibly affect your flock. The old adage that an ounce of prevention is worth a pound of cure certainly applies to growing turkeys.

It is best to raise different types and ages of poultry separately. In addition, do not allow unnecessary visitation to your flock and do not visit other poultry flocks. Early detection of most diseases will aid in their treatment.

Preventive measures include a good sanitation program and a vaccination program designed to protect the flock from any diseases that may be prevalent in your area. A vaccination program should not be considered a substitute for good management.

Risk Factors for Disease Transmission

- Infected birds within a flock
- Newly acquired birds added to an existing flock, especially birds coming from shows or fairs
- Different species of birds reared together or in close quarters
- Different ages of the same species reared together or in close proximity
- New, and especially young, birds reared in a previous flock that had not been properly cleaned and disinfected
- Eggs or poults from infected breeders
- Humans; hands, hair, feet or shoes, and clothes can harbor infectious agents
- Wild birds, rodents, flies, darkling beetles, other insects, and parasites
- Contaminated feed, water, or air
- Contaminated vaccines and medications
- Contaminated equipment brought onto the farm, such as trucks, tractors, coops, and egg flats
- Vaccines that are so potent that they cause the disease rather than prevent it

If you purchase stock from a good, clean source, follow a sound sanitation program, use a good feeding program, and provide a comfortable growing environment, you have gone a long way toward keeping your flock healthy. However, losses do occasionally occur. Commercial flock owners, for example, expect a mortality rate somewhere between 8 and 15 percent. So if you lose one bird and the rest of the flock is eating and drinking and looks healthy, don't get too excited about it. However, a disease in the flock is usually accompanied by several warning signs: (1) a drop in feed or water consumption, (2) the appearance of sick or dead birds, and (3) a change in the birds' behavior and appearance. When it is apparent that a disease is present, seek the advice of a trained poultry diagnostician. Do not use drugs or antibiotics indiscriminately; this can do more harm than good, and the only result may be a waste of money.

If there are no local diagnosticians, you may submit sample birds to a state diagnostic laboratory. The sample should include two or more sick, or recently dead, birds. Preserve dead specimens by keeping them cool to prevent decomposition. Early diagnosis and fast treatment are always recommended as the quickest ways to solve poultry disease problems. The addresses of the state diagnostic laboratories can be found in the back of this book. Alternatively, you can contact your local county Cooperative Extension Service by looking in the phone book under County Government. Ask for a poultry or livestock agent. This person can help you contact the state diagnostic laboratory and can address many management issues.

Turkey Diseases

Several diseases and parasites may affect turkeys. Only the more common ones are described here, and these are not discussed in great detail. For more in-depth information, you can consult many excellent texts on poultry diseases (see References). Also, discuss poultry disease issues with some of your rural veterinarians. If you can find one with an interest in birds, especially poultry, she can be of invaluable assistance. This would be especially true if the veterinarian is consulted in conjunction with a state diagnostic laboratory.

Recognizing a disease problem at the onset, diagnosing it before it becomes widespread in the flock, and getting treatment started early can greatly reduce the possible losses due to mortality, morbidity, and diminished overall performance.

Aspergillosis (Brooder Pneumonia)

Aspergillosis is usually a disease of young birds, but it can affect older birds, too. The symptoms include the following:

- The birds stop eating.
- Breathing may be rapid.
- The birds may gasp and have labored breathing.
- Eyes may be inflamed.
- Eyelids may swell and stick together.

This disease is caused by a fungus that is inhaled by the birds and usually comes from moldy litter or feed. On the postmortem examination, yellow-green nodules may be found on the lungs and in the trachea, bronchi, and viscera. There is no known treatment.

Prevention

Spread of infection may be prevented by culling the sick birds, thoroughly cleaning and disinfecting the house and equipment, and carefully removing moldy litter or feed from the building.

Avian Influenza

There are more than one hundred influenza-virus isolates from birds. The majority are from ducks and turkeys. The disease can be mild or acute. The mild form produces listlessness, respiratory distress, and diarrhea; the acute form causes air sacculitis and sinusitis with cheesy exudates. Large drops in egg production can occur. The best treatment is prevention. Avian influenza seems to become a problem when husbandry and sanitation are below par. Keep wild birds, especially migratory waterfowl, away from turkey flocks. The risks are even greater if you keep waterfowl that attract wild birds during the spring and fall migration periods.

If avian influenza is known to be a problem in an area, it is well to keep in mind that the organism can be transmitted in many ways from one farm to another. For example, it can be transmitted on clothes, equipment, egg boxes, and poultry crates.

Blackhead (Histomoniasis)

Blackhead is caused by the protozoan parasite *Histomonas meleagridis*. It affects turkeys of all ages. It can also affect chickens; however, the disease tends to cause less mortality in those birds. Since chickens may act as an intermediate host for the organism that causes blackhead, it is recommended that they not be kept in the same house and never be intermingled with turkeys. Ideally, they should not be kept on the same farm. The term *blackhead* is somewhat misleading because that sign may or may not be present with the disease.

Cecal worm eggs can harbor for long periods the organism that causes blackhead. When picked up by the turkeys, it infects the intestines and liver. Both chickens and turkeys can host the cecal worm.

Mortality with this disease may reach 50 percent if treatment is not started and the infection checked immediately.

The signs of this disorder include drooping heads, dark heads, and brownish-colored, foamy droppings. On necropsy, inflammation of the intestine and ulcers on the liver may be seen.

Prevention

Incidence and severity of the disease depends on the management and sanitation programs used. Several measures are helpful in preventing blackhead:

- Follow good sanitation practices in the brooding facilities
- Rotate the range areas
- Segregate young birds from old birds
- Separate turkeys from chicken flocks

Coccidiosis

Coccidiosis is a common disease of poultry and is caused by *Coccidia*, a group of protozoan parasites. The birds become exposed by picking up sporulated oocysts in fecal matter and litter. It should be assumed that all flocks grown on litter or range are disposed to the disease. Birds raised on elevated wire or slats are not exposed to droppings and normally don't contract coccidiosis. However, if feces are retained in the pen or contaminate the feed or water, even birds on wire or slats can be placed at risk for the disease.

Coccidia are host specific; that is, the coccidia that affect turkeys do not affect chickens. Different species of the parasite affect different parts of the digestive tract. Six species are known to infect turkeys, but only three of them are commonly troublesome. If left unchecked, the disease can be fatal.

Signs of Coccidiosis

You should suspect coccidiosis in your flock if you notice the following:

- Ruffled feathers
- Head drawn back into the shoulders and the appearance of being chilled (birds having this appearance are sometimes called *unthrifty*)
- Bloody diarrhea

Necropsy findings may include lesions and hemorrhages in various parts of the intestine, depending on the particular species of parasite.

Treatment and Prevention

The disease can be treated through use of sulfonamides or other coccidiostats as prescribed by a diagnostician or service person.

Coccidiosis may be prevented or controlled by feeding coccidiostats at low levels in the starter feed.

Erysipelas

Erysipelas, which means red skin, is caused by the bacterium *Erysipelothrix insidiosa*. Swine, sheep, humans, and other species are also susceptible to the disease.

The signs of erysipelas are swollen snoods, bluish purple areas on the skin, congestion of the liver and spleen, listlessness, swollen joints, and yellow-green diarrhea.

Erysipelas is primarily a disease of toms because the organism readily enters through wounds caused by fighting. Since the snood is frequently injured when toms fight, this is a common site for erysipelas infection. For this reason, some commercial producers have their turkeys' snoods removed at the hatchery or on the farm upon arrival. Erysipelas is a soil-borne disease, and contaminated premises are the primary source of infection.

Treatment and Prevention

The disease responds well to penicillin, and tetracycline is also effective. However, a veterinarian should be consulted for treatment.

Control requires good management and sanitation. Vaccination is recommended for areas in which the disease is common. If this disease is suspected, use care. Wear gloves when performing a necropsy on a diseased bird.

Fowl Cholera

Fowl cholera is caused by the bacterium *Pasteurella multocida*. This disease is highly infectious and affects all domestic birds, including turkeys. The birds become sick rapidly and may die suddenly without showing signs. When signs do appear, they include listlessness; fever; excessive consumption of water; diarrhea; swelling of the head and face sinuses in the chronic form; red spots or hemorrhages on the surface of the heart, lungs, or intestines, or in the fatty tissues on postmortem examination; and swollen liver (that is, the liver has a cooked appearance with white spots).

Treatment and Prevention

Treatment with sulfonamides, such as sulfaquinoxaline and sulfamethazine, is currently recommended. Sulfaquinoxaline in the feed at 0.33 percent for 14 days is considered to be one of the best treatments. Antibiotics are sometimes injected at high levels.

Good management practices are essential to prevention. Sanitary conditions in the poultry house, range rotation, and proper disposal of dead birds help to prevent cholera. In problem areas, vaccines can be used and are recommended.

Fowl Pox

Fowl pox is found in many areas of the United States. It is caused by a virus and is spread through contact with infected birds or by such vectors as mosquitoes and other biting insects or wild birds. There are two forms of fowl pox — the dry, or skin, type and the wet, or throat, type.

Birds with fowl pox have a poor appetite and look sick. The wet pox causes difficult breathing; nasal or eye discharge; and yellowish, soft cankers of the mouth and tongue. The dry pox causes small, grayish white lumps on the face. These lumps eventually turn dark brown and become scabs. On postmortem examination, cankers may be found in the membranes of the mouth, throat, and windpipe. There may be occasional lung involvement or cloudy air sacs.

Although the disease has no treatment, antibiotics may help to reduce the stress associated with it. The only means of control is by vaccination, which is recommended in areas where fowl pox is a problem.

Mycoplasma-*Related Diseases*

Mycoplasma bacteria may cause several types of disease conditions in turkeys and other species of birds. There are several strains of *Mycoplasma* bacteria but the ones of primary concern are *Mycoplasma gallisepticum*, *Mycoplasma synoviae*, *Mycoplasma iowae*, and *Mycoplasma meleagridis*. Outbreaks of disease caused by these organisms result in a variety of symptoms and bring about poorer growth rates and egg production, along with possible flock morbidity and mortality.

Mycoplasma organisms are extremely small compared to bacteria and do not have a rigid cell wall. These organisms can survive for up to several days outside the bird on feathers, clothes, and hair, for example. Once a flock is infected with *Mycoplasma*, the best course of action is to depopulate the farm and to clean and disinfect everything. Have a down time of at least 2 weeks and then restart production.

Infectious Sinusitis

Infectious sinusitis is a disease of turkeys caused by M. *gallisepticum* — the same organisms that cause chronic respiratory disease in chickens. The disease is also found in pigeons, quail, pheasants, ducks, and geese. These bacteria are transmitted through the egg from carrier hens. Stress is thought to lower the poult's resistance to the disease (this tends to be true for most diseases).

Affected birds show nasal discharge, coughing, difficulty breathing, foamy secretions in the eyes, swollen sinuses, decreased feed consumption, and weight loss. Air sac infection may be in evidence on postmortem examination.

Treatment and Prevention. Antibiotics in the feed or water are useful to help control *Mycoplasma* infections. Individual treatment with injectable penicillin and streptomycin in the sinuses can also be useful. Obtaining poults from *Mycoplasma*-free breeding stock is the most important aspect of disease control.

Infectious Synovitis

Synovitis is an infectious disease of turkeys caused by M. *synoviae*. It was first identified as a cause of infections of the joints, but more recently it has been shown to cause respiratory disease as well. This disorder can affect birds of all ages. The bird species mentioned in the M. *gallisepticum* section are also susceptible to M. *synoviae*.

Infectious synovitis causes lameness, reluctance to move, swollen joints and foot pads, weight loss, and breast blisters. Some flocks have respiratory symptoms. Greenish diarrhea occurs in dying birds.

The most common means of transmission of synovitis is through infected breeders. Poor sanitation and management practices also contribute to the problem.

Postmortem findings include swelling of the joints; presence of a yellow exudate, especially in the hock, wing, and foot joints; possible signs of dehydration; enlarged liver and spleen; and air sacs filled with liquid exudate. Aside from findings on necropsy, respiratory involvement is not easy to spot.

Treatment and Prevention. Antibiotics yield some results, and they should be given by injection or in the drinking water. Some producers prefer to give antibiotics by both methods simultaneously for the best results. Always obtain poults from *Mycoplasma*-free breeders.

Mycoplasma iowae Infection

M. *iowae* has been shown to be responsible for reduced hatchability in turkeys. It is transmitted through the egg from the breeder hen like the other types of mycoplasma. It can be lethal to turkey embryos. The disease is best prevented by obtaining poults from M. *iowae*–free breeder flocks.

Mycoplasma meleagridis Infection

Like the other mycoplasma disorders, M. *meleagridis* is an infectious disease of turkeys that a breeder hen transmits to the egg.

The main sign is air sacculitis. Even though this type of infection is thought to be specific to turkeys, it may occur in peafowl, quail, and pigeons. Obtain poults from *M. meleagridis*–free stock.

Newcastle Disease

Acute and highly contagious, Newcastle disease is a respiratory disorder that is caused by a virus and is found in chickens, turkeys, and other species of poultry. It causes high mortality in young flocks. In breeder flocks, egg production frequently drops to zero. Newcastle spreads rapidly through the flock.

Signs of the disease are gasping, coughing, hoarse chirping, increased water consumption, loss of appetite, huddling, partial or complete paralysis of the legs and wings, and holding of the head between the legs or on the back with neck twisted. Postmortem examination may reveal congestion and hemorrhages in the gizzard, intestine, and proventriculus; cloudy air sacs may also be noted.

The disease is transmitted in many ways: It can be tracked in by people or brought in by birds from another site, dirty equipment, feed bags, or wild birds. There is no effective treatment, though antibiotics are normally given to limit secondary invaders.

Prevention

Vaccination is recommended in most areas of the country and can be administered to an individual bird or on a mass basis. On an individual basis, the birds can be vaccinated intranasally, ocularly, or in the wing web. (The wing web is the thin layer of skin at the forward edge of the wing between the proximal or shoulder end of the humerus and the tip of the wing.) On a mass basis, the vaccine can be given to the birds in drinking water or in the form of a mist or spray. Follow the manufacturer's recommendations when using these products, and conform to the vaccination program that is recommended for your area.

Omphalitis

Omphalitis is caused by a bacterial infection of the navel and occurs when the navel doesn't close properly after hatching. It can also be caused by poor sanitation in the incubator or hatchery, chilling, or overheating.

Signs of omphalitis may include weakness; unthriftiness; huddling; an enlarged, soft, mushy abdomen; and an infected navel surrounded by a bluish black area.

Mortality may be high for the first 4 or 5 days of life. There is no treatment for the disease. Most of the affected poults die within the first few days, and no medication is needed for the survivors.

Salmonella-Related Diseases

More than two thousand species of the genus *Salmonella* have been identified. Although quite a number of these species can affect chicken and turkeys under certain conditions, very few are a serious threat to the poultry industry. Of greatest concern are *Salmonella pullorum*, *Salmonella gallinarium*, *Salmonella arizonae*, and paratyphoid infection. Paratyphoid is caused by many species of *Salmonella* and can infect animals as well as poultry.

Pullorum

S. pullorum is an infectious disease of chickens, turkeys, and some other species and is found all over the world. The National Poultry Improvement Plan was organized in 1935 by the U.S. Department of Agriculture to eradicate pullorum as well as fowl typhoid. Pullorum causes high mortality, which most often occurs at 5 to 7 days of age. Pullorum is sometimes called white diarrhea.

Birds with pullorum appear droopy, huddle together, act chilled, and may have diarrhea and pasting of the vent.

S. pullorum is transmitted from the hen to the poult mainly through the egg. After transmission, it spreads rapidly through the down of poults located in incubators and hatchers.

One of several types of blood tests can help establish a positive diagnosis for *S. pullorum*. Flocks that have pullorum should be depopulated or destroyed immediately and definitely not kept as replacements or for breeding purposes. Buy poults from pullorum-free hatcheries only.

Postmortem Findings. Examination of diseased chicks has revealed dead tissue in the heart, liver, lungs, and other organs and an unabsorbed yolk sac. The heart muscle may be enlarged and have grayish white nodules. The liver may also be enlarged, appear yellowish green, and be coated with exudate.

Fowl Typhoid

Fowl typhoid is caused by the bacterium *S. gallinarum*. It affects chickens, turkeys, and other species of birds and may be present wherever poultry is grown.

Affected birds may look ruffled, droopy, and unthrifty and have a loss of appetite, increased thirst, and yellowish green diarrhea. Postmortem examination may show a mahogany-colored liver, an enlarged spleen, and pinpoint necrosis in the liver and other organs

Prevention. Typhoid is prevented in the same manner as pullorum: Buy typhoid-free poults. Flocks that are positive for fowl typhoid should be destroyed.

Arizona

S. arizonae causes an infectious disease that can affect chickens but most commonly strikes turkeys. This disease, which is also called paracolon infection, has both acute and chronic forms. Many serotypes of the disorder occur in mammals, birds, and reptiles. Mortality usually occurs in the first 3 to 4 weeks of life. There can be high *morbidity* (that is, sickness) without high mortality.

The disease has no distinct signs; however, the following symptoms may occur: unthriftiness, blindness, infections of the intestinal tract, peritonitis, and mottled and enlarged livers on necropsy.

Diagnosis is based on laboratory isolation of the organism. Various drugs are used to minimize mortality from this form of salmonella in poults. The disease is most commonly transmitted by hen to egg to poult, but it can also be spread by direct contact with infected birds, rodents, and contaminated premises.

Prevention. Blood testing of breeders and ensuring proper sanitation of the hatchery and other environments are important for prevention. Good rodent control is imperative for control of this and other salmonelloses.

Paratyphoid

Paratyphoid is an infectious disease of turkeys and some other birds and animals. It is caused by one or more of the *Salmonella* bacteria other than those discussed in the preceding sections. Transmission may be from the hen through the egg to the chick. The organism is also found in fecal matter of infected birds.

The disease primarily infects young birds but may also affect older birds. In young birds, mortality can run as high as 100 percent.

Some birds may die of paratyphoid without showing signs. However, you may notice weakness, loss of appetite, diarrhea, and pasted vents. Birds may appear chilled and huddle together for warmth. Older birds lose weight, are weak, and have diarrhea.

On necropsy, birds that have died of paratyphoid reveal unabsorbed yolk sacs, small white areas on the liver, inflammation of the intestinal tract, congestion of the lungs, and enlarged livers. Older birds may have white areas on the liver, but most typically show no lesions.

Prevention. Some antibiotics may reduce losses, prevent secondary invading organisms, and increase the bird's appetite. The disease can be controlled through sanitation and isolation of the flock from sources of infection, such as wild birds, birds from other flocks, rodents, and contaminated feed and equipment.

Turkey Coronavirus (Bluecomb)

Turkey coronavirus is a highly contagious disease of turkeys of all ages.

Signs of coronavirus include depression, subnormal body temperature, diarrhea, loss of appetite, weight loss, poor growth, poor feed conversion, watery feces, dehydration, and prostration.

Some flocks of turkeys with coronavirus seem to be healthy and show few signs of the disease. However, flocks that test positive for this virus usually do not perform as well (with respect to growth and feed conversion) as flocks that test negative, even when they do not show signs. Mortality can be very low or extremely high in poults but is usually low in older birds. Coronavirus is spread by direct or indirect contact with infected birds or contaminated premises.

Prevention

Some antibiotics may help in cases of high mortality with secondary bacterial problems. However, good husbandry and management and strict adherence to good biosecurity practices are the best prevention for coronavirus.

Reducing the Chance of Reinfection with Coronavirus

If a flock tests positive for coronavirus, do the following:

1. Depopulate the farm.
2. Clean and disinfect everything.
3. Let the facilities sit empty for 4 to 6 weeks.

Be careful not to spread litter from infected flocks around other turkey or poultry flocks. Some turkey producers have observed that letting litter sit undisturbed for 2 weeks before removal from the turkey house cuts down on transmission.

Turkey Parasites

Several parasites — both internal and external — can affect poultry, but relatively few of them are of major importance.

Internal Parasites

Some internal parasites may cause setbacks in weight gain and a loss of egg production in laying birds; severe infestation can cause death. Some intestinal parasites harbor other disease organisms that may be harmful to turkeys. Good management and the type of management system used play important roles in the control of internal parasite infestations.

Large Roundworm

Light infestations of roundworms are probably not a cause for concern. However, when the worms become numerous, birds can become unthrifty and feed conversion and weight gain suffer. These worms may also reduce egg production. By themselves, the worms infrequently cause mortality, but they can cause fatalities if they occur along with other diseases.

Causes of Roundworm Infestation

The following environmental factors can contribute to roundworm infection:

- Ranges that have been used for several years
- Houses with dirt floors
- Houses that haven't been properly cleaned and disinfected

Turkeys become infected with roundworms by picking up the eggs from feces or contaminated ranges or quarters.

The large roundworm is 1½ to 3 inches (3.8–7.5 cm) long. It is found in the upper to middle portion of the small intestine. If the birds are heavily infested, the worms may extend for the full length of the small intestine. Piperazine and other wormer compounds are used to treat birds with roundworms and can be given in the water, the feed, or a capsule.

Clean, dry litter aids in the control of roundworm infestations in the growing houses. Where turkeys are free-ranged, regular rotation of the range area has been found to be quite effective in controlling worm infestations.

Cecal Worms

Cecal worms are very small and by themselves are not injurious. However, they are significant because they act as carriers for the organism that causes blackhead.

Gapeworm

Gapeworms attack the bronchi and trachea and can cause pneumonia, gasping for breath, and even suffocation. Small turkeys open their mouths with a gaping movement and may have bloody saliva. Mortality may be high among young infected birds. Gapeworms can quickly build up drug resistance; therefore, treatment should be tailored for each farm or flock. The earthworm is an intermediate host. The gapeworm is fork-shaped and red in color.

Tapeworm

There are several species of tapeworms, varying in size from microscopic to 6 to 7 inches (15–17.5 cm) in length. They are flat, white, and segmented and inhabit the small intestine. They cause weight loss and lowered egg production. Tapeworms need intermediate hosts like worms, snails, or beetles to complete part of their life cycle. Turkeys get tapeworms by eating the infected worms, snails, or beetles.

Control of Internal Parasites

Effective control of internal parasites depends primarily on a program of cleanliness and sanitation. Parasite eggs can remain viable in the soil for more than a year. This means that it's important to rotate poultry runs or yards. Preferably, poultry ranges should be used for 1 year and left idle for 3 before they are used again. Poultry yards and runs should be located in well-drained areas and be kept as clean as possible. Cultivating and seeding down these areas helps prevent the birds from picking up parasite eggs.

External Parasites

Although there are many external parasites of poultry, few are of major importance. However, certain external parasites, especially when present in large numbers, can cause loss in weight or loss of egg production, as well as decreased growth rates and feed efficiency. Only a few of the more important external parasites are discussed here. Birds should be handled and closely observed on a regular basis to catch external parasite infestations early.

Lice

Lice are chewing and biting insects that cause birds considerable grief. With severe infestations, growth and feed efficiency suffer; lice can also affect egg production. They irritate the skin and result in scab formation.

Lice spend their entire lives on the birds and die within a few hours if separated from the host. The eggs *(nits)* are laid on the feathers, where they are held with a gluelike substance. The eggs hatch in a few days to 2 weeks. Lice live on the scale of the skin and feathers. Several types attack poultry. Lice can be gray or yellow; it's difficult to distinguish between the colors. The body louse, one of the most common poultry lice, usually affects older birds. The lice and their eggs are seen on the fluff, the breast, under the wings, and on the back.

Treatment. Insecticides that may be used to treat lice are carbaryl, malathion, coumaphos, and pyrethrins or permethrin. Treat the birds according to the directions on the label, and examine them frequently for signs of reinfestation.

Mites

There are a number of species of mites that are capable of influencing flock performance. Some live on the birds; others spend more time off the birds. Mites' mouths are adapted either to chew or to pierce. They live on blood, tissue, or feathers, depending on the type. Generally, mites cause irritation and affect growth and egg production. In the case of severe infestations of certain mites, they can cause morbidity, or even mortality, in the flocks.

Monitor for Mites

Mites can be particularly troublesome, in respect to both their effect on turkeys and the effort needed to eradicate them. Keep a close and watchful eye on your birds to catch mite infestations early so that treatment will be effective.

The Northern Fowl Mite. Northern fowl mites are a reddish, dark brown. These mites are found around the vent, tail, and breast and live on the birds at all times. They attach to feathers and suck blood, causing anemia, weight loss, and reduced egg production. Materials recommended for treatment of northern fowl mites are carbaryl, malathion, and pyrethrins (permethrins). Use according to the manufacturer's directions.

The Chicken Mite. Also known as red mites, chicken mites feed at night and are not found on the birds during the day. During daylight hours, they may be seen on the underside of roosts, in cracks in the wall, or in seams of the roosts. Other signs are salt-and-pepper-like trails under roost perches and clumps of manure. Red mites are bloodsuckers and cause irritation, weight loss, reduced egg production, and anemia. Treat the chicken mite with the same insecticides as those used for the northern fowl mite.

Precautions for Drug and Pesticide Use

To maintain a healthy flock and to obtain optimum production, it is sometimes necessary to administer drugs or pesticides. Use all drugs and pesticides according to the label directions, and use them with caution. *Never* use a pesticide that is not registered for use on poultry.

For treatment of any disease, but especially if drug administration is necessary, you should involve a veterinarian. Although it may take effort to locate a local veterinarian who treats birds, it will be worth the effort. Also, be sure to utilize your state livestock diagnostic services. For more information, contact a poultry or livestock Extension agent at your local Cooperative Extension Service. Look in the phone book under Local or County Government.

Disease Prevention

There is no substitute for good management. Prevention is worth a lot more than treatment. Drugs or pesticides are not intended as substitutes — they work best when combined with good sanitation and sound management practices. Early diagnosis and treatment of a disease or parasite problem is important.

Probiotics

An alternative to antibiotics is a class of feed additives called probiotics. These are fermentation products that contain either live cultures of beneficial bacteria or by-products of fermentation, such as mannan oligosaccharides (MOS). The term *saccharide* refers to sugars, and MOS products are complex sugars that are not digested by the animal. Instead, they attach to pathogenic bacteria and prevent them from adhering to the gut wall. The animal's digestive systems can then remove the bound bacteria from its system. Probiotics are just beginning to gain favor with poultry producers. However, if you want to avoid antibiotics but wish to provide some protection for your turkeys, probiotics may be worth a try.

Drug Withdrawal Periods and Tolerance

Federal agencies have established withdrawal periods and tolerance levels for various agents used in poultry production. For example, the U.S. Food and Drug Administration (FDA) insists that certain drugs be withdrawn a specified number of days prior to slaughter, and some pesticides cannot be used within a certain number of days. These withdrawal periods vary from 1 or 2 days to several days and are subject to change. The FDA also establishes maximum amounts of residues for certain chemicals. Some insecticides can be used around poultry but not directly on the birds, on the eggs, or in the nests. Frequency of use may also be restricted for some agents.

Because withdrawal periods and tolerances and accepted forms of treatment do change, specific precautions for various agents are not addressed here. The point is that drugs and insecticides must be used discriminately. Follow all precautions on the label. If used improperly, drugs and insecticides can be injurious to humans, animals, and plants.

Chemical Safety Measures

Keep drugs and pesticides in their original containers in a locked storage area, well out of reach of children and animals. Avoid inhaling sprays or dusts, wear protective clothing, use recommended equipment, and be *safe.*

Nutritional Deficiencies

A number of poultry disorders may be caused by nutritional deficiencies or imbalances. With today's well-formulated diets, nutritional problems occur infrequently, so a thorough discussion of nutritional deficiencies will not be undertaken here. However, to underscore the importance of good nutrition, a few of the more common nutritional problems will be mentioned.

Rickets

Rickets is caused by a deficiency of vitamin D_3, phosphorus, or calcium or by a calcium–phosphorus imbalance. It may occur in birds on range or when grains are used along with complete feeds or protein concentrates because the birds may not get enough calcium. Provide oyster shell to these birds.

Birds with rickets exhibit weakness; stiff, swollen joints; soft beaks; soft leg bones; and enlarged ribs.

Perosis

Perosis in turkeys, sometimes called *slipped tendon,* is a leg problem typically caused by a deficiency of choline in the diet. Heredity may also be a factor. Other dietary factors include biotin, folic acid, manganese, zinc, and possibly pyridoxine. Swollen hocks can also be effected by deficiencies of niacin and vitamin E. In perosis, the large tendon of the leg at the rear of the hock slips to one side, resulting in a twisted leg. If permanently crippled, the bird should be killed. Most turkeys with perosis respond to early use of additional manganese in the feed.

Miscellaneous Problems

Some problems do not easily fit into the aforementioned categories but do need to be described. They will probably be easily observed when they occur in your flock.

Good management will prevent some of the miscellaneous problems. However, on occasion a few birds in a flock develop abnormalities and have to be removed from the flock. Leg problems, for example, may make it difficult for the affected birds to get to feeders and waterers. Sometimes these birds can be segregated from the flock and nursed back to health, but frequently they have to be sacrificed in a humane manner. This type of mortality, plus normal mortality, requires a sanitary bird-disposal program. Rodents are also a common problem on most poultry farms, making a rodent-control program necessary.

Pendulous Crop

In the normal position, the crop is in the wishbone cavity and is attached to the side and back of the neck. If for some reason the connective tissues that hold the crop in place weaken, the crop drops. If the crop gets too far out of its normal position, feed cannot pass from the crop to the gizzard and the bird actually starves with the crop full of feed. Young birds with a mild pendulous crop condition may recover. Seriously affected birds seldom recover, and treatment is ineffective.

Leg Weaknesses

Leg weaknesses other than perosis may be caused by vitamin deficiencies or by such diseases as infectious synovitis. In day-old poults, a condition resembling perosis is called *spraddle legs*. This disorder may be due to a genetic factor, faulty incubation, or a deficiency in the diet of breeding stock. Smooth, slippery surfaces in hatching trays, shipping boxes, or under brooders that cause poor footing may also result in spraddle legs. Place young poults on wire, paper with a rough surface (if paper is used), or litter to avoid spraddle legs. Another leg problem, crooked toes, may be hereditary or due to faulty brooding conditions.

Euthanasia

Anyone who decides to raise or keep livestock should consider how to humanely kill sick and crippled animals that need to be removed from the group. If you are raising turkeys for slaughter, you might take euthanasia in stride. However, if you are keeping turkeys as a hobby flock in which the birds reach a near-pet status, the situation may be quite different and even traumatic. Even in hobby flocks, there are times when birds should be humanely killed. For owners of such flock or people who view their birds as pets, the use of a veterinarian might be best advised. However, producers can perform euthanasia for their birds when needed. The options for euthanasia include blunt trauma, cervical dislocation, carbon dioxide overdose, and drugs.

Blunt Trauma

In this context, *blunt trauma* is the striking of the head with a heavy, blunt instrument, 6 to 8 inches (15–20 cm) in length. This may sound crude and objectionable to some; however, it can be very effective and humane if done properly. Turkeys do not have thick skulls, and a heavy blow with an appropriate object kills the bird immediately. The turkey should be restrained so that the blow is properly placed to kill the bird rather than to cause injury. Blunt trauma is usually considered for older animals but is effective for young birds as well.

Cervical Dislocation

This method of dispatch involves dislocation of the neck vertebrae from the cranium. However, it is important to realize that cervical dislocation includes the separation of the spinal cord and carotid arteries. When done properly, the bird is killed instantly.

Carbon Dioxide (CO_2) Overdose

Another method of humane death, the carbon dioxide chamber can be simply a box with a lid on the side or top. To construct the chamber and dispatch a bird:

1. Drill a hole in the box for a hose from a CO_2 tank.
2. Place the bird in the box.
3. Close the lid.
4. Turn on the gas.

Within several minutes, the bird calmly loses consciousness and dies.

Warning!

If CO_2 overdose is your preferred method of dispatch, note that a box built for a large bird would also be large enough for a small child. An overdose of CO_2 kills a person as readily as it does a turkey. For this reason, construct small CO_2 boxes for use on small groups of young birds.

Drug Overdoses

Any drug overdose must be administered by a veterinarian. This is a humane method that may suit hobby and pet bird owners best, but it can be expensive for even modest-size flocks. (For more information on euthanasia of turkeys, contact the National Turkey Federation.)

Disposal of Dead Birds

A good biosecurity program includes proper disposal of dead birds. There shouldn't be much mortality in small-flock production, but dead birds should still be disposed of properly. Some of the best and most practical methods are burial, incineration, and composting.

Burial is convenient and inexpensive. However, be sure to check with the state veterinarian or Extension Service to see if burial is legal. Also, be sure to follow the guidelines that the state or Extension Service might provide. Keeping away wild or feral animals from buried poultry or any other buried livestock is very important. Burial may not be allowed in some states or portions of states where the water table is relatively high.

Incineration is also convenient but is more expensive than burial. You must purchase an incinerator as well as the fuel to operate it. Properly incinerated poultry carcasses do not have an offensive odor when the flame or part of the flame is located between the carcasses and the exhaust pipe. Incineration is a biosecure method of carcass disposal.

Composting is an old and natural phenomenon that has recently been used to dispose of dead poultry. Composting involves more labor than burial or incineration; however, it is a biosecure and environmentally friendly method of disposal. Composting does not involve rotting carcasses, and if done properly, there is no offensive odor.

If the flock is large enough to generate a significant number of dead birds, a compost bin can be constructed. Make this bin small if you plan to empty it with a shovel, or about 6 feet by 6 feet (1.9 x 1.9 m) or larger if you're using a tractor and bucket. More information on composting can be obtained from your local Cooperative Extension Service office.

How to Compost Birds

This natural method of carcass disposal is simple.

1. Layer turkey carcasses with straw and feces or with used or old turkey litter containing feces. The carcasses and the feces contain bacteria that degrade the carcasses.
2. Make sure that no carcasses are directly on the ground or exposed to the air; the bottom and top layer should be litter or straw.
3. Add some water, but do not use so much that you drench the material or have water running out the bottom; however, too little moisture results in dried cadavers or mummies.
4. Keep the compost pile out of the weather, and avoid direct exposure to wind or rain.
5. Composting is a heat-generating operation. Once the material has heated and then cooled, the pile can be turned and the material goes through another cycle.

A successful process results in material that looks similar to used turkey litter. There may be a few long bones left, especially if older birds are composted.

Rodent Control

A successful rodent-control program consists of three equally important phases: exclusion, habitat modification, and poison baiting.

Exclusion prevents rodents from entering the premises or building. This is difficult but worth the effort in helping to keep down rodent numbers. To exclude rodents, seal all holes and openings with concrete, sheet metal, or heavy-gauge hardware cloth.

Habitat modification involves changing the environment of the building or premises to one that is less suited to rodents. The goal is to eliminate sources of food, water, and shelter for rodents.

- Eliminate food sources by cleaning and removing all feed spills, keeping all feed storage areas neat and clean, cleaning feeders and removing spoiled feed, and removing empty feed bags.
- Eliminate water sources by checking all waterers and plumbing for leaks and removing any containers that collect rainwater.
- Eliminate shelter by removing unused equipment, including feeders, waterers, and garbage cans, and removing all vegetation from around buildings; either mow closely or kill.

Poison baiting is the only phase of rodent control that rapidly and consistently reduces rodent populations. However, its effectiveness is reduced when the exclusion and habitat phases of the program are not done efficiently. Commercially available baits contain grains mixed with one of several types of rodenticides.

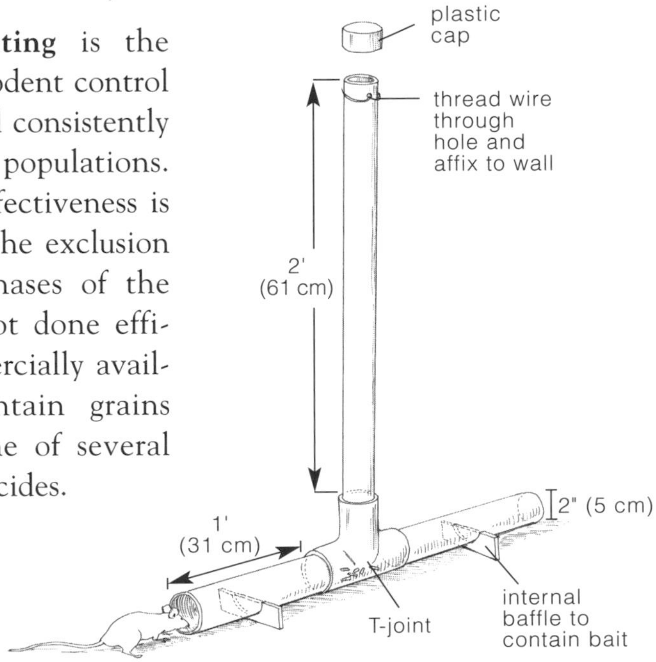

This rodent bait station is easily constructed from PVC pipe. The cap at top is kept loose. Remove the cap and pour in the bait. The rodents eat in seclusion, then die elsewhere.

Single- and Multiple-Dose Poisons

Rodenticides are classified into two groups: single dose and multiple dose. Single-dose poisons quickly knock down rodent numbers but may cause bait shyness. Bait shyness occurs when a rodent does not return to the bait after consuming a nonlethal dose, thereby reducing effectiveness. Multiple-dose baits are usually anticoagulants and require repeated ingestion during a 4- to 5-day period to be lethal. Because they are slow acting, multiple-dose poisons do not normally cause bait shyness. Some of the more potent anticoagulants, such as bromadiolone and brodifacoum, cause death after a single feeding.

Bait stations are essential to a successful rodent-control program because they provide a sheltered, secure place for rodents to feed; protect bait from dust, moisture, and weather; allow regular inspection of the bait to make sure rodents are using it; and keep other animals and children out of the bait

Other methods of rodent control include traps, glue boards, fumigants, and tracking powders, but none of these is as practical for controlling rodents as baiting.

Points to Remember about Baiting

Maintain a constant supply of clean, dry poisoned bait. Check daily and remove any wet or soiled bait. Bait should be maintained in the best possible condition to ensure acceptance by rodents. In many cases, an effective rodent-control program can save you money in feed alone.

SIX

KILLING AND PROCESSING

Turkeys should be finished and ready for processing at 12 to 24 weeks of age for old varieties and 12 to 18 weeks of age for newer, heavier varieties. Hens are usually processed at younger ages than toms. Turkey broilers or fryer-roasters are usually animals of the same strain used to produce heavier carcasses but are processed at younger ages, such as 8 to 10 weeks of age. The precise age for finishing and processing depends on the turkey variety and strain, the feeding program, and other factors.

Assessing Readiness for Processing

To assess whether a bird is in prime condition and ready to be processed, see if it is free of pinfeathers. The bird is "ready" when the feathers are easy to remove. Pinfeathers are immature feathers that do not protrude or may have just pierced the skin. Short protruding feathers have the appearance of a quill with no plume. They are unattractive, particularly in varieties with dark feathers, and cause downgrading when present in finished market birds. If the bird is not going to be marketed but, rather, consumed at home, the pinfeathers may stay in place; however, if the presence of pinfeathers is considered to be a drawback, it is best to delay dressing those birds until the feathering improves.

You must also check the degree of fat covering.

Evaluating Degree of Fat Covering

1. Pull a few feathers from the thinly feathered area of the breast, at a point about halfway between the front end of the breastbone and the base of the wing.

2. Take a fold of skin between the thumb and forefinger of each hand.

3. Examine for thickness and coloration. On a prime turkey, the skin fold is white or yellowish white and quite thick. Well-fattened birds have thick, cream-colored skin, while underfattened birds have thin (often paper-thin) skin that is semitransparent and tends to be reddish.

Care Before Killing

Careless handling can cause birds to pile and trample each other, resulting in injuries. Recent injuries may appear red at the bruise site; old injuries are bluish green. Such defects detract from the dressed appearance. Always catch birds properly.

Withhold feed from the birds for approximately 10 hours before killing; however, do not withdraw water or excessive dehydration may occur. Removing the feed enables the crop and intestines to empty before killing and makes the job of eviscerating much cleaner and easier. Remove the birds to be feed-restricted from the pen, and put them into coops containing wire or slat bottoms to keep them away from feed, litter, feathers, and manure. After catching the birds, keep them in a comfortable, well-ventilated place prior to killing. Overheating or lack of oxygen can cause poor bleeding and result in bluish, discolored carcasses.

Proper Bird-Catching Technique

1. Grab the legs between the feet and hock joints with one hand.
2. Straighten the legs to lock the hock joints. Don't grasp the legs at the feathered area above the hock joints, as this may cause skin discoloration.
3. After catching the bird by the shanks, hold one wing at the base with the other hand. This immobilizes the bird effectively. It also gives the handler control of the bird and prevents injuries and bruising. ▶

Processing Area and Equipment

Home processing of just a few birds requires little in the way of special facilities or equipment; but if a fairly large number of turkeys are to be dressed, you should have an adequate area and some special equipment, such as a mechanical picker.

Process your poultry in as sanitary a manner as possible. It is important to prevent contamination of the carcasses. One of the most common sources of contamination is the contents of the intestine. Contamination, which can also come from dirty facilities, equipment, or people, reduces quality and shelf life — that is, the period before spoilage begins.

Getting Started

Plan to process the turkey in a clean, well-lighted area that has a water supply and no flies. It is helpful to have flat surfaces that can be easily cleaned, and suitable containers for handling the *offal* (or waste by-products).

At best, the processing job is a messy one. Ideally, there should be two rooms available for processing. If several birds are to be done at one time, use one room for killing and plucking the birds and the other for finishing, eviscerating, and packaging. If this is not possible, or just a small number of birds is involved, do the killing and plucking in one operation, clean the room, and then draw and package the birds as a second operation. When you can use only one room, following this method makes the procedure far more sanitary. Good organization makes the process go more smoothly.

Ideally, the processing equipment should be made of metal or other impervious material to facilitate plant cleaning and sanitation. The processing plant requires a plentiful water supply, at least 5.5 to 10 gallons (21–38 L) of water per dressed turkey.

Shackles or Killing Cones

If only a few birds are to be dressed, a shackle for hanging can be made from a strong cord with a block of wood, 2 x 2 inches (5.1 x 5.1 cm) square, attached to the lower end. A half hitch is made around both legs and the bird is suspended upside down. The block will prevent the cord from pulling through.

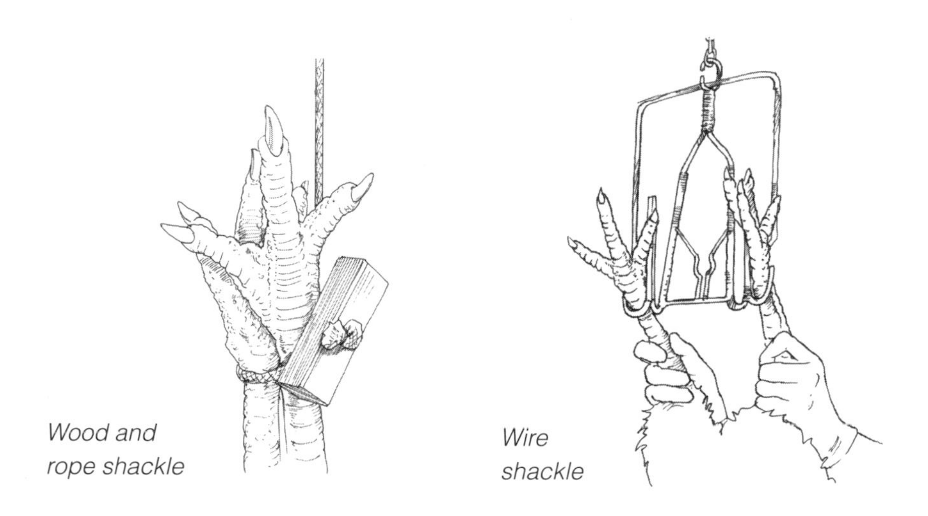

Wood and rope shackle

Wire shackle

Commercial and semicommercial dressing plants use metal shackles that hold the legs apart and allow easy plucking. Some producers make their own shackles out of heavy-gauge wire. Other people prefer to use killing cones, which are similar to funnels. The bird is put into the cone with its head protruding through the lower end. This restrains the bird and reduces struggling, which can lead to bruising or broken bones.

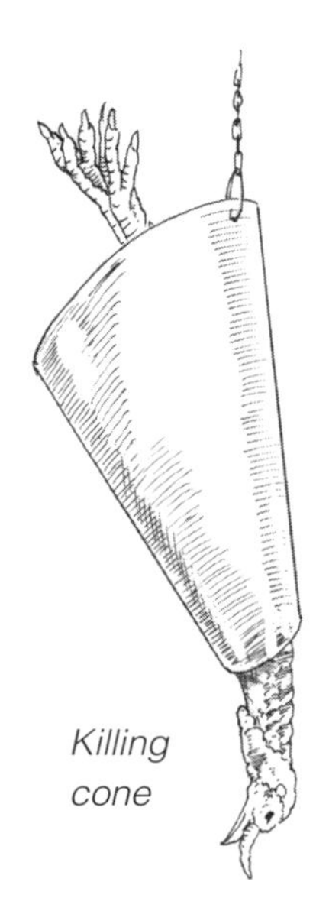

Killing cone

Weights

A weighted blood cup or a simple weight attached to the lower beak of the bird prevents it from struggling and splashing blood. You can make the device from a window weight attached with a sharp hook to the lower beak. The blood cup is not used when killing cones are available.

You can make a blood cup from a 2-quart (1.9 L) can. Solder a sharp-pointed, heavy wire to the can. The wire hooks through the lower beak. Weight the cup with concrete or heavy stones.

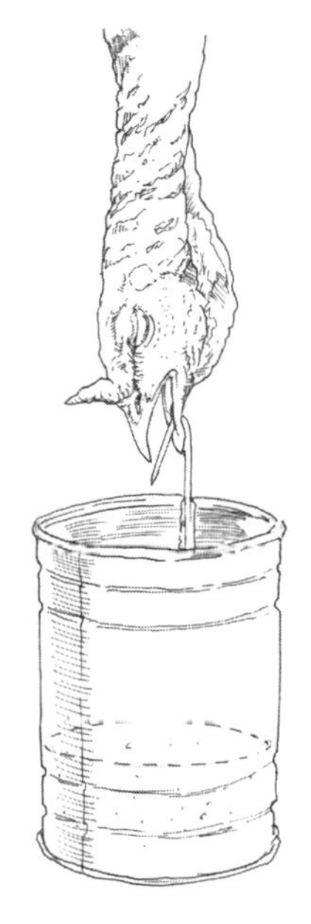

A homemade blood cup (not needed with a killing cone)

Knives

Just about any type of knife can be used for dressing poultry. There are special knives for killing, boning, and pinning. Six-inch (15 cm) boning knives work well. If the birds are to be brained, then use a thin sticking or killing knife. Make sure all knives, especially the killing knives, are very sharp.

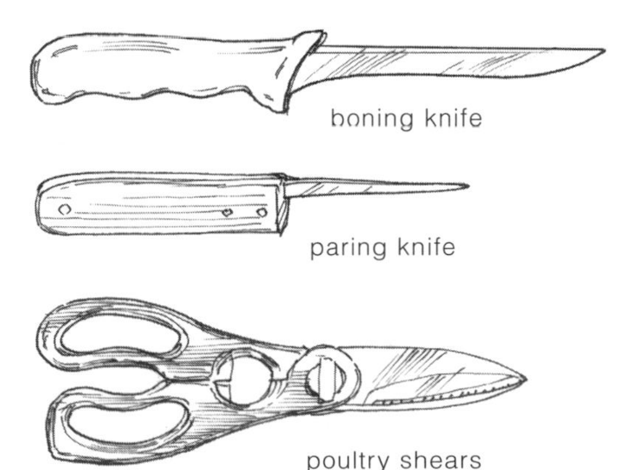

Common poultry knives and implements

Scalding Tank

When the birds are to be scalded and only a few birds are to be dressed, a 10- to 20-gallon (37.9–75.7 L) garbage can, or any other clean container of suitable size, is satisfactory. When a considerable amount of dressing is being done, a thermostatically controlled scalding vat is preferred. In the absence of the automatically controlled vat, hot water can be continually heated and the vat replenished as required to maintain a desired temperature.

Homemade, electrically heated scalding vat with thermostatic control

Thermometer

Accurate temperatures are important for scalding. Acquire a good, rugged dairy thermometer, a candy thermometer, or a floating thermometer that accurately registers temperatures between 120° and 150°F (48.9° and 65.6°C).

Killing

There are many methods of killing turkeys. The first method that follows is perhaps the simplest. An alternate, slightly more difficult method is also provided. Both involve severing the bird's jugular vein. The jugular vein needs to be thoroughly severed to ensure that the birds are well bled. With either method, make sure the killing knife is razor sharp, which will allow for a more humane kill. A bird that is not well bled will have a purplish skin color that seriously affects the bird's dressed appearance and marketability.

Method 1

1. Suspend the turkey by its feet with a rope or metal shackle, or place it in a killing cone.
2. Hold the head with one hand and pull it down to exert slight tension, which steadies the bird.
3. With a sharp knife, sever the jugular vein just behind the mandibles. This can be done by inserting the knife into the neck close to the neck bone, turning the knife outward, and severing the jugular. It may also be done by cutting from the outside.

Method 2

The jugular vein can also be severed from inside the mouth; this is slightly more difficult than the previous method.

1. Hold the head in one hand, with your fingers grasping the sides of the neck, taking care not to squeeze the jugular vein.
2. Make a strong, deep cut across the throat from the outside close to the head so that both branches of the jugular vein are severed cleanly at or close to the junction. *Warning:* Be sure to hold the head so your fingers do not get in the way. ▶
3. Do not grasp the wings or legs to the extent that you restrict blood flow from these parts. Incomplete bleeding results in a poor-appearing carcass.

Birds can be slaughtered either conscious or unconscious. Combination stunning and killing knives are frequently used. The knife has an electrical component with a button. The knife is held next to the bird's head, and the bird is stunned when the button or switch is on. The stunning renders the bird unconscious. The switch is turned off, and the bird is slaughtered.

Debraining

Debraining loosens the feathers so that it is easier to pluck the birds. It is done after the jugular vein is cut in birds that are to be dry-picked, but it may also be done when the carcasses are to be scalded (see below) or to make feather removal even easier. Though dry-picking is slower, the outer layer of skin is not removed, making for a fine-appearing dressed carcass.

How to Debrain

This procedure requires considerable practice before proficiency is achieved.

1. Insert the knife through the groove or cleft in the roof of the mouth.
2. Push the knife through to the rear of the skull so that it pierces the rear lobe of the brain as shown. ▶
3. Rotate the knife in a one-quarter turn. This kills the bird and loosens the feathers.

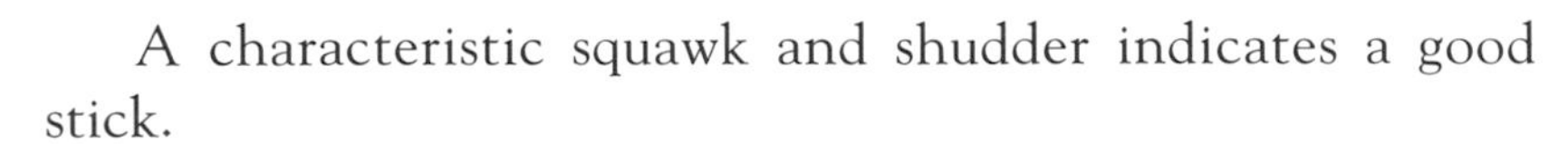

A characteristic squawk and shudder indicates a good stick.

Scalding

There are two methods of scalding: subscalding and semiscalding. Both work equally well.

Subscalding

As soon as the bird is dead and bleeding is complete (usually 2 to 3 minutes), loosen the feathers using the subscald method. Dunk the bird in water at approximately 140°F (60.9°C) for about 30 seconds. The subscald method makes it easy to remove the feathers and gives the skin a uniform color. The skin surface tends to be moist and sticky and will discolor if not kept wet and covered. For the scald to be effective, slosh the bird up and down in the water to get the water around the follicles at the base of the feathers.

Semiscalding

Another method that is sometimes used is semiscalding. The bird is scalded for 30 to 60 seconds in water 125° to 130°F (51.7°–54.4°C). With the semiscald method, the feathers loosen but the temperature is not hot enough to destroy the outside layer, or skin cuticle. Thus, the carcasses look more like dry-picked birds.

Water Temperature and Timing Are Critical

For an effective semiscald, the water temperature must be maintained within the narrow range of 125° to 130°F (51.7°–54.4°C). Time is also a factor and varies with the age of the bird. If the water is a little cool or the scalding time too short, the feathers will not loosen enough for easy picking. If the feathers are difficult to pull out, skin tears can result. If the water is too hot or the scalding time is too long, the bird will have an overscalded or patchy appearance.

Plucking

If available, a rubber-fingered plucking machine can remove the feathers as well as the *cuticle* (or bloom), which is the thin, outer layer of the skin. Remaining pinfeathers are removed by hand. Don't let the skin dry out or it will become discolored. If they are not immediately eviscerated, put the birds in cold running water.

Hand-Plucking

Hand-pluck feathers this way:

1. Rehang the bird on the shackle.
2. With a twisting motion, remove the large wing and tail feathers first.
3. Remove the remainder of the feathers as quickly as possible in small bunches to avoid tearing the skin.

Pinning and Singeing

Pinfeathers, the tiny, immature feathers, are best removed under a slow stream of cold tap water. Use slight pressure and a rubbing motion. You can use a pinning knife or a dull knife to pluck the feathers that are difficult to remove. By applying pressure between the knife and the thumb, you can squeeze out the pinfeathers. The most difficult may have to be pulled. Usually, turkeys have a few hairlike feathers left after they have been hand-plucked. You can singe these hairs with an open flame. A small gas torch works well. Do not apply the flame directly on the carcass to avoid scorching the skin.

The pinning and singeing process may sound time consuming, but it will improve the appearance of the carcass as well as increase customer demand for your product.

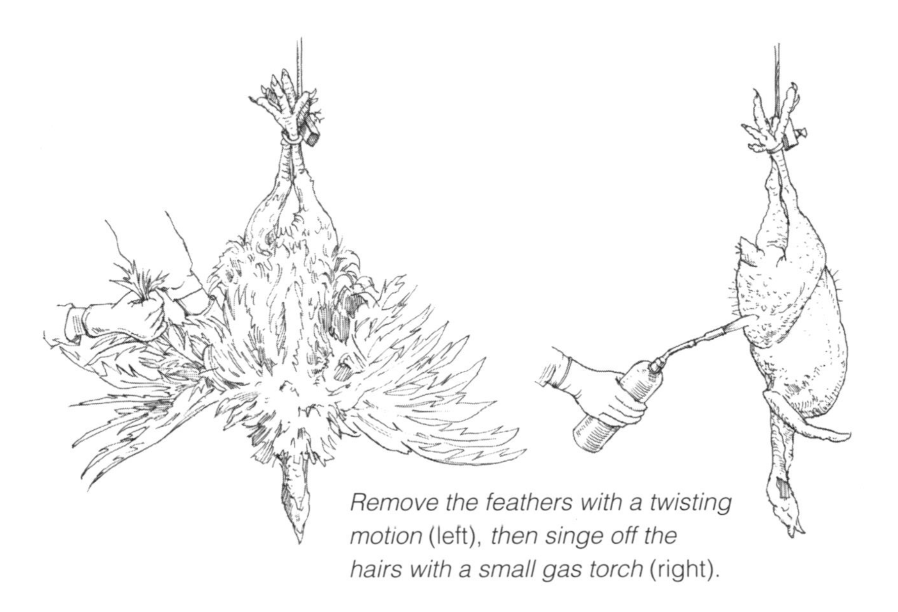

Remove the feathers with a twisting motion (left), *then singe off the hairs with a small gas torch* (right).

Eviscerating

After picking and singeing, wash the carcasses in clean, cool water. They are ready for evisceration as soon as they are washed. Some prefer to cool the poultry first because, after cooling, eviscerating is somewhat easier and cleaner. Others eviscerate and then place the birds in ice water or cool water that is constantly replenished. There are many methods of eviscerating poultry, but the most important part of the process is to keep your working area and equipment clean.

The parts of the turkey are removed in the following order: (1) tendons (optional); (2) shanks and feet; (3) preen gland (oil sac); (4) crop, windpipe, gullet, and neck; (5) lungs, liver, and heart attachments; (6) lungs, gonads, and kidneys.

Wear Gloves

Practice good sanitation when handling and processing the carcass — wear disposable surgical, latex, or vinyl gloves.

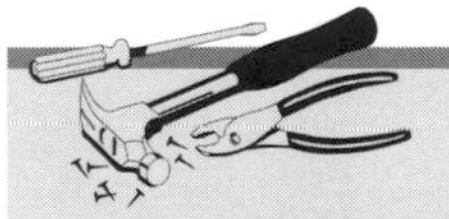

MATERIALS

Only a few items are needed for evisceration:

- Sharp, stiff-bladed boning knife
- Hook (if the leg tendons are to be pulled)
- Solid block or bench on which to work
- A piece of heavy parchment paper or meat paper that is laid on the working surface and changed as necessary

Removing the Tendons, Shanks and Feet, and Oil Sac

Sometimes the tendons are removed from the drumsticks before removing the shanks and feet. Removal of the tendons makes carving and eating the drumsticks easier. By cutting the skin along the shank, the tendons that extend through the back of the leg may be exposed and twisted out with a hook or a special tendon puller, if available.

Remove the oil sac on the back near the tail, as it sometimes gives the meat a peculiar flavor. This is removed with a wedge-shaped cut.

When cutting off shanks and feet at the hock, leave a flap of skin behind the hock.

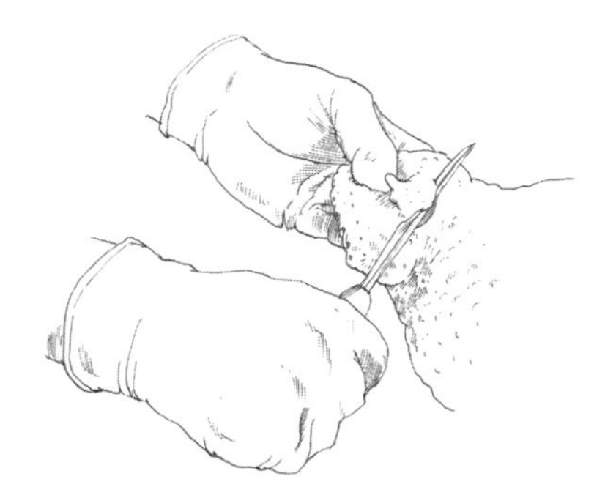

Remove the preen gland (oil sac) near the tail.

Removing the Crop, Windpipe, Gullet, and Neck

1. To remove the crop, first cut off the head.

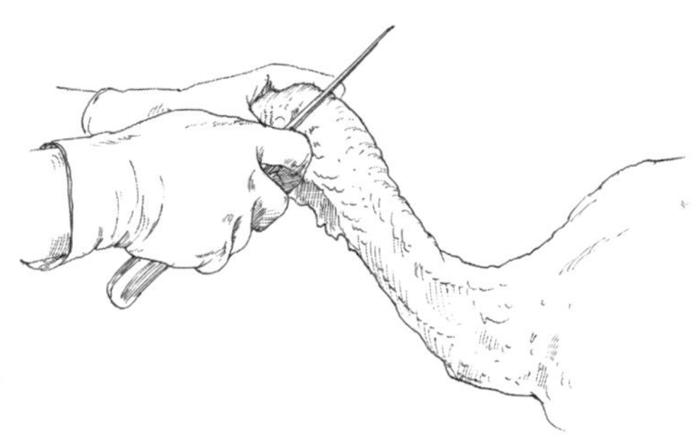

2. Slit the skin down the back of the neck to a point between the wings.

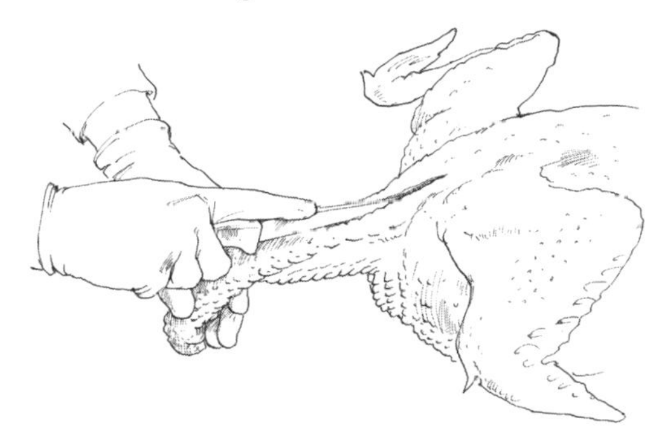

3. Separate the skin from the neck and then from the gullet and windpipe.

4. Follow the gullet to the crop and remove, being careful to cut below it.

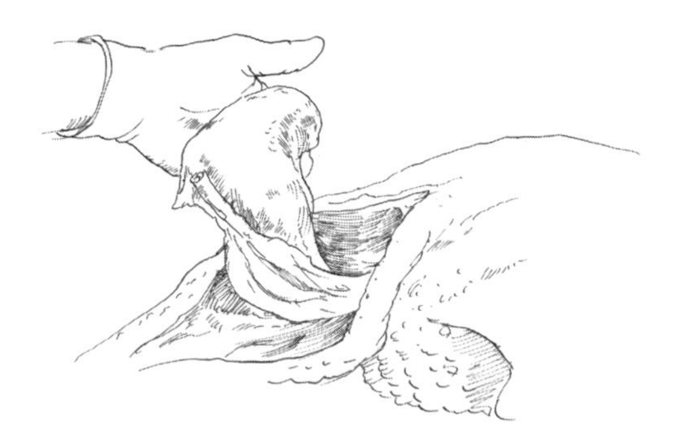

5. Cut off the neck as close to the shoulders as possible. A pair of heavy shears is handy for this purpose; you can also sever the neck by cutting around its base with a knife and then breaking and removing with a twisting motion.

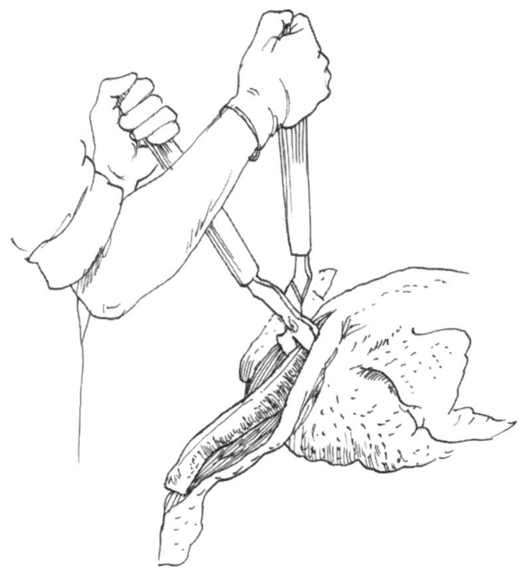

6. Loosen the vent by making a circular cut around it. Do this carefully to avoid cutting into the intestines.

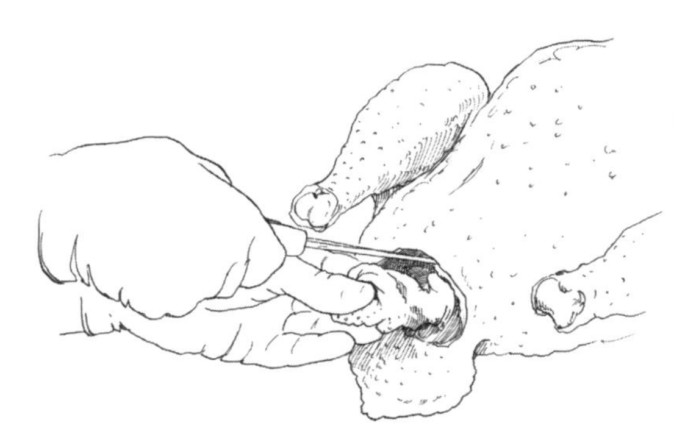

Removing Abdominal Contents

1. Make a short, horizontal cut 1½ to 2 inches (3.8–5.1 cm) between the vent and the tip of the keel bone; make the horizontal cut about 3 inches (7.6 cm) long.

2. Break the lungs, liver, and heart attachments carefully by inserting the hand through the rear opening.
3. Loosen the intestines by working the fingers around them and breaking the tissues that hold them.
4. Remove the viscera through the rear opening in one mass by hooking two fingers over the gizzard, cupping the hand, and using a gentle pulling and slight twisting motion.

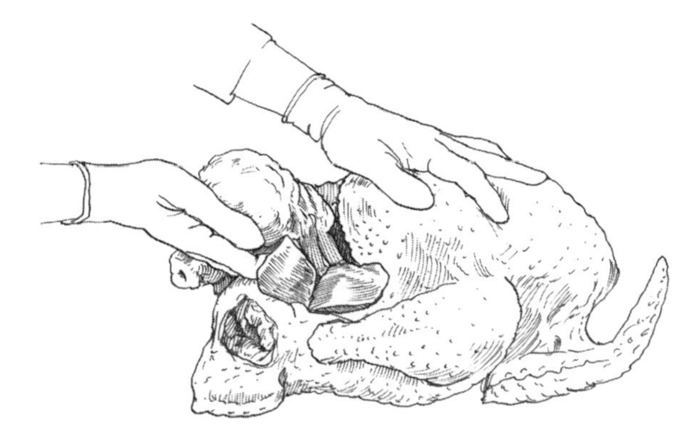

5. Remove the lungs, gonads, lungs, and kidneys. The lungs are attached to the ribs on the either side of the backbone. These can be removed by using the index finger to break the tissues attaching them to the ribs. Insert a finger between the ribs and scrape the lungs loose. The lungs appear pink and spongy. The gonads are also attached to the backbone.

Washing the Carcass

1. Wash the inside of the carcass using water from a faucet or from a clean hose.
2. Wash the outside to remove any adhering dirt, loose skin, pinfeathers, blood, or singed hairs.
3. Hang the bird to drain the water from the body cavity.

Cleaning the Giblets

Remove the gallbladder, which is the green sac attached to the liver, without breaking it. If the gallbladder breaks during removal of the viscera or while cleaning the liver, the bile is likely to give a bitter, unpleasant taste to any part it contacts and will cause a green discoloration.

If you are careful, a cool gizzard can be cleaned without breaking the inner lining. Cut carefully through the thick muscle until a light streak is observed. Do not cut into the inner sac or the gizzard lining. The gizzard muscle may then be pulled apart with the thumbs, and the sac and its contents will be removed unbroken, if you're lucky.

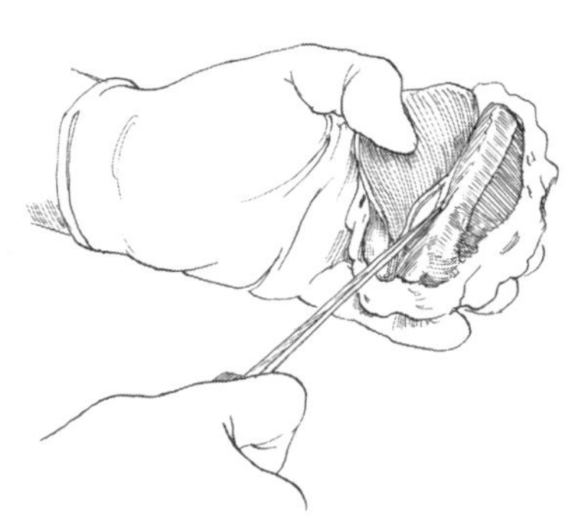

Cut through the gizzard to the light streak.

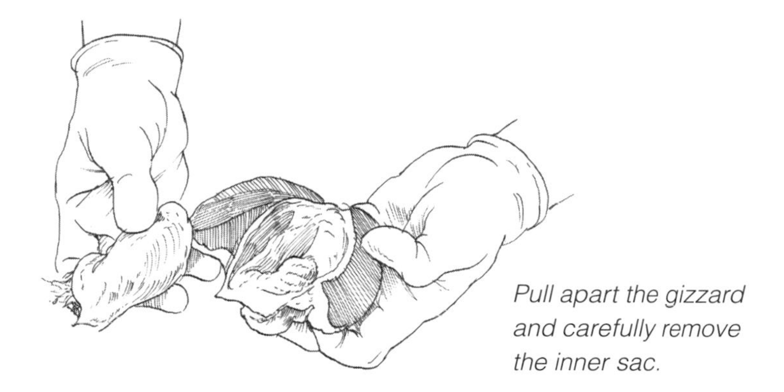

Pull apart the gizzard and carefully remove the inner sac.

Trussing

A properly trussed bird appears neat when packaged. Proper trussing also conserves juices and flavors during roasting. The simplest method to truss a bird follows:

1. Tuck the hock joints under the strip of skin between the vent opening and the cut from which the viscera were removed.

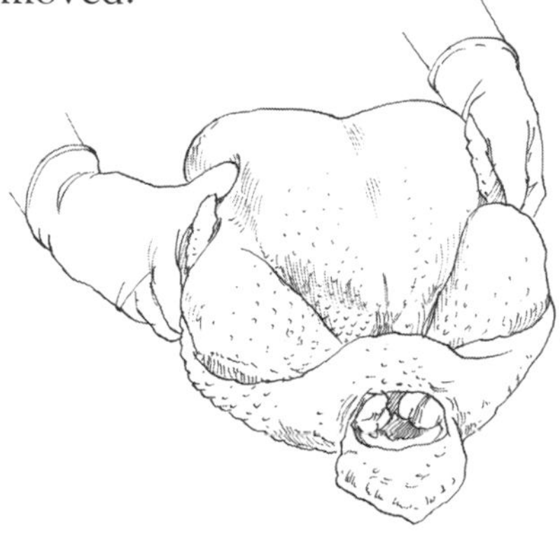

2. Although turkeys are usually packaged with the wings in a natural position, another option is to draw back the neck flap between the shoulders and fold the wing tips over the shoulders to hold the skin in place.

Chilling and Packaging

The next phase of processing your turkeys involves chilling the birds in water; wrapping the giblets; and wrapping the turkey.

It is important to cool the birds as soon as possible after killing. If cooling is done slowly, bacteria can develop and cause spoilage and undesirable flavors.

Guidelines for Air-Cooling

If birds are to be air-cooled, the air temperature should be from 30° to 35°F (-1°–1.7°C). The time required to cool the carcasses depends on the size of the bird and the temperature of the air. Birds to be air-cooled should always be packaged to avoid discoloration.

You can cool poultry with water if air-cooling is not possible. If dressed with excessively high, scalding temperatures or for too long a period, the skin of air-cooled birds may be blotchy and discolored. When scalding temperatures are too high, water-cooling is the preferred method of cooling the carcasses. Dressed birds may be cooled in tanks of ice water or under cold running water. The important factor is to maintain a constant temperature of 34° to 40°F (1°–4.4°C). For a bird's internal temperature to become that cool, it must remain in the water for 5 to 10 hours, again depending on the size of the carcass. If eaten or frozen immediately after dressing, carcasses tend to be tougher than if aged for a while.

Remove the carcasses from the water and hang them up to dry for 10 to 30 minutes before packaging. Make every effort to remove all of the water from the body cavity before putting a bird in the bag.

Wrapping the Giblets

Spoiled giblets can spoil the entire carcass, so always wrap them well. Use this method to ensure proper packaging:

1. Wrap the giblets — that is, the neck, gizzard, heart, and liver — in a sheet of wax paper or a small plastic bag.
2. Stuff the giblets into the body cavity or under the neck skin.

Bags for Wrapping Poultry

There are two types of bags available for poultry. One is the so-called Cryovac bag (W.R. Grace and Co., Duncan, South Carolina); the other is a common plastic bag. When the turkey is placed in the Cryovac bag and then boiled, the bag shrinks and adheres to the bird. Not only do these bags make a nice-appearing package, but they also help to reduce the amount of water that builds up during the freezing process. Good-quality plastic bags are also available and do a satisfactory job of maintaining quality in frozen, dressed poultry. The bags should be impermeable to moisture to prevent dehydration during freezing, which causes toughness.

After chilling, insert the carcass in a plastic bag. Suck out the air with a vacuum cleaner or a plastic hose (left) *and secure the bag with a twist tie* (right).

Birds to be bagged should be trussed thoroughly (see above), then inserted front end first into the plastic bag. After the bird is in the bag, you can remove excess air by using a vacuum cleaner or by inserting a flexible hose into the top of the bag and then creating a vacuum. Merely keep the bag snug around the hose or vacuum cleaner, then suck the air out of the bag. Twist the bag several times and secure it with a twist tie or a rubber band.

Fresh-dressed, ready-to-cook turkeys have a shelf life of approximately 10 days if refrigerated at a temperature of 29° to 34°F (-1.7°–1°C). If you plan to freeze your turkey, do it by the third day after it is dressed and chilled. Chill the poultry to below 40°F (4.4°C) before placing it in the freezer.

The amount of weight a turkey loses between slaughter and dressing varies by age and the type.

Dressing Percentage — Live to Eviscerated Weight*

Type of Bird	Live Weight (pounds)	Blood and Feather Dressed (percentage lost)	Eviscerated (percentage lost)
Broilers and fryers	5–6	7	25
Hens — small	10–12	6	18
Hens — medium	12–15	6	20
Hens — large	18–20	7	20
Toms — small	20–25	7	20
Toms — medium	25–30	7.5	20
Toms — large	35–40	7.5	19

* 1 lb = 0.454 kg.

Source: Estimated using method reported by J. Brake et al. Relationship of sex, strain, and body weight to carcass yield and offal production in turkeys. *Poultry Science* 24:161–168, 1995.

State and Federal Grading and Inspection

Some processors of poultry that is sold off the farm are subject to the Poultry Products Inspection Act. There are exemptions for small producers, and regulations may vary among states. Check the regulations that apply in your area. For information on grading and inspection programs and how they affect you, contact your state department of agriculture.

SEVEN

PRESERVING, COOKING, AND SERVING TURKEY

In the previous chapter, we discussed the slaughter and processing of turkeys in detail — from live to oven-ready, packaged in a freezer bag. Turkeys can be held for up to 10 days at about 32°F (0°C). If they are to be frozen, they should be frozen by the third or fourth day.

Preserving

Freezing is probably the simplest and most popular method of preserving the turkey. If stored in a moisture-resistant bag, turkey can maintain its quality for many months. Loss of moisture during freezing and storage causes a drier, tougher bird when cooked. Usually, oven-ready turkeys are packaged in heavy plastic bags (see chapter 6).

Blast-freezing (moving air) at temperatures of -20° to -35°F (-28.9°–-37.2°C) freezes the carcass quickly and makes a better-quality frozen product. After the birds are blast-frozen, store them at -5°F to -20°F (-20.6°–-28.9°C).

The quality of fresh-frozen turkeys is similar to that of fresh turkeys, especially if directions for proper handling and thawing are followed. More on this later.

Smoked-Cooked Turkey

Smoked-cooked turkey is a delicacy. This turkey product has been on the market for many years, particularly during the holiday season. Some stores carry it year-round as a specialty item. The smoking-cooking process is relatively simple and can be done without expensive ingredients or equipment.

Making Salt Brine for a Smoked-Cooked Turkey

Several brine preparations are available. Salt brine can be prepared at home. For 50 pounds (22.7 kg) of turkey, use:

- Water, 10 gallons (37.9 L)
- Salt (noniodized), 8 pounds (12½ cups [3.6 kg])
- Brown sugar, 3 pounds (7½ cups packed [1.4 kg])
- Saltpeter, 1 cup (236.6 mL)

If your water is chlorinated, boil it for 15 minutes to get out the chlorine, which may affect the curing reaction. Salt brines are corrosive to many metal containers so use a large jar, crock, wooden barrel, stainless-steel container, or plastic tub. Just add the ingredients to the water and stir until dissolved. The temperature of the brine should be 40°F (4.4°C).

Preparing and Smoking the Turkey

A more intense flavor can be obtained if the turkeys are injected with the brine solution. A syringe with a No. 12 needle works well.

1. Inject brine deeply into the thickest part of the breast, thighs, drumsticks, and joints. Inject an amount equal to approximately 10 percent of the weight of the turkey.
2. Immerse the turkey in the brine for 3 days.
3. Move the turkey around in the brine at least once during the curing process to make sure the brine penetrates the entire carcass.
4. Remove the carcass from the brine, rinse in cold water, drain, and place on the rack in a smoke-cooker. For a more traditional-looking product, wrap in stockinettes or cheesecloth before putting in the smoker.
5. Hang or place the bird in a water smoker-cooker. This type of cooker has one or more grills on which the meat can be placed for cooking. The grills are placed over a pan of water that is heated by charcoal or by a propane gas vent in the bottom. ▶ 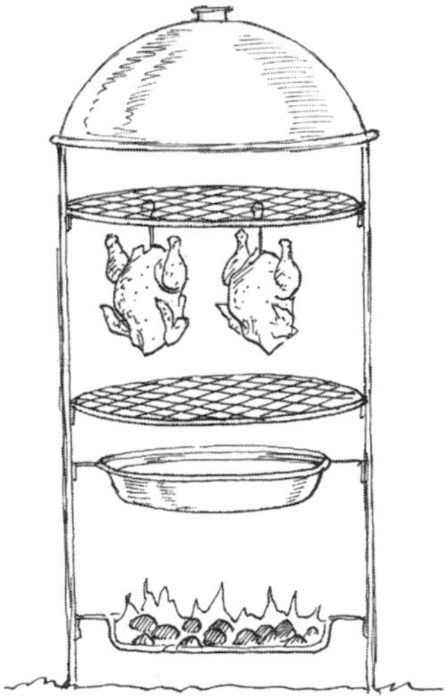
6. A good flavor is obtained by using hickory, maple, apple chips, or other hardwood chips. Soak the chips in water before using. Moist chips are used to prevent flare-ups and to prolong the smoke for a longer period. Place the moist chips directly on the heat source or in a metal pan placed on the heat source.

The meat is cooked when the internal temperature of the thickest part of the breast muscle reaches 170°F (76.7°C).

Recipe for Smoking Smaller Portions

Another recipe for curing and smoking smaller amounts of turkey has been developed by the National Turkey Federation. Do the following to prepare for curing and smoking according to the federation's methods:

1. Place the turkey in a sufficiently large plastic container and pour water over the turkey until it is covered by about 2 inches (5.1 cm).
2. Remove the turkey.
3. Measure the amount of water needed to cover the turkey and add salt, sugar, and saltpeter to water in the proportions given in the chart (see below) for the amount of water used.
4. Grind spices in a blender and add to brine, stirring vigorously until salt, sugar, and saltpeter are dissolved.
5. Place the turkey in the brine and allow it to cure in the refrigerator for 2 to 4 days.
6. Remove the turkey and dry it with paper towels before placing it on a rack or rotisserie over hot coals.
7. For smoking, add a few soaked hickory or fruitwood chips every half hour.
8. Keep the grill or rotisserie covered to keep smoke in; add more charcoal as needed.
9. When the meat thermometer registers 170°F (76.7°C), the turkey is cooked and ready to eat.

Curing and smoking are not methods for preserving a turkey; it must *always* be refrigerated. Several types of smoke-cookers are available commercially.

Ingredients for Brine Solution (approximately 10%)

Water	1 gallon	2 gallons	3 gallons	4 gallons	5 gallons	6 gallons	7 gallons
Salt	1 cup	2½ cups	4 cups	5½ cups	7 cups	8½ cups	10 cups
Sugar	⅓ cup	⅔ cup	1 cup	1⅓ cups	1⅔ cups	2 cups	2⅔ cups
Saltpeter	2⅔ tsp.	5⅓ tsp.	8 tsp.	10⅔ tsp.	13⅓ tsp.	16 tsp.	18⅔ tsp.
Bay leaves	3	6	9	12	15	18	21
Coriander seeds	3	6	9	12	15	18	21
Whole cloves	4	8	12	16	20	24	28
Whole peppercorns	8	16	24	32	40	48	56

1 gallon = 3.8 L; 1 cup = 236.6 mL; 1 teaspoon = 4.9 mL.

Thawing a Frozen Turkey

If the turkey is frozen, leave it in the original bag and use one of the following methods to thaw it:

- To thaw slowly, place the turkey on a tray in the refrigerator for 24 hours for each 5 pounds (2.3 kg) of turkey.
- If you're in a hurry, immerse the turkey in cold water while it's still in the watertight bag. Change the water occasionally so it remains cold. This method requires about ½ hour per pound (0.5 kg) of turkey.

Refrigerate or cook the turkey as soon as it is thawed. If you plan to stuff it, do so just before cooking. Refreezing an uncooked turkey is not recommended.

Preparation for Roasting

After thawing the turkey, do the following:

1. Take the turkey out of the plastic bag.
2. Remove the neck and the giblets from the body cavity; these can be cooked for broth for flavoring dressing and for giblet gravy.
3. Rinse the turkey and wipe it dry.
4. If the turkey is to be stuffed, stuff it loosely. Allow ¾ cup (177.5 mL) of stuffing per pound (0.5 kg) of oven-ready weight.
5. If the turkey is not being stuffed, rub salt in the cavities and, if desired, put in a few pieces of celery, carrots, onion, and parsley for flavor.
6. Tie down the legs, or tuck them in the skin flap. The neck skin can be skewered to the back and the wing tips folded back under, in toward the body.

Roasting a Turkey

1. Place the turkey, breast up, on a rack in a shallow pan.
2. Brush the carcass with butter, margarine, or cooking oil if desired.
3. If using a meat thermometer, insert it into the thickest part of the thigh or breast, making sure the thermometer doesn't touch the bone.
4. Roast in an oven set at 325°F (162.8°C). When the thermometer in the thigh registers 180°F (82.2°C) or the thermometer in the breast reaches 170°F (76.7°C), the bird is done.

Foil placed loosely over the turkey eliminates the need for basting, but basting is fine if you prefer. Remove the foil during the last half hour to allow the turkey to brown well.

When the stuffing reaches a temperature of 160° to 165°F (71.1°–73.9°C), it is thoroughly cooked. With large birds, it may be difficult to reach this temperature, so if the bird weighs 24 pounds (10.9 kg) or more, you should probably cook the stuffing separately.

Using a meat thermometer is the most accurate way to determine when the turkey is done. Some people say turkey is done if the drumstick feels soft when pressed with the thumb and forefinger or when it moves easily.

Approximate Roasting Time for Turkey in Preheated 325°F (162.8°C) Oven*

READY-TO-COOK WEIGHT	APPROXIMATE COOKING TIME
8–12 pounds (3.6–5.4 kg)	3–3½ hours
12–14 pounds (5.4–6.4 kg)	3½–4 hours
14–18 pounds (6.4–8.2 kg)	4–4¼ hours
18–20 pounds (8.2–9.1 kg)	4¼–4¾ hours
20–24 pounds (9.1–10.9 kg)	4¾–5¼ hours

*Times given are approximate. Because turkeys may vary in individual conformation, the exact degree of thawing is difficult to determine, and individual ovens vary in temperature (an oven set at 325°F [162.8°C] can range from 300° to 350° [148.9°–176.7°C]). The size of the bird and whether it is stuffed also affect the roasting time.

Note: Roast unstuffed turkey ½ hour less than times given.

Recipes

Roasted turkey conjures fond memories of Thanksgiving dinners for many people, but turkey can be served all year long. The possibilities are endless. There are many fine recipes for preparing gravies and stuffings, and some of the best are included here. The gravy recipe that follows is extremely popular. Vary the recipe, if you wish, by adding giblets. The stuffing and dressing recipes that follow are favorites in different regions of the country. All are excellent recipes that complement the delicious flavor of roasted turkey.

Good Gravy!

When the turkey is done, you are ready to make the gravy.

1. Pour drippings from the roasting pan into a bowl, leaving all of the brown particles in the pan.
2. Let the fat rise to the top of the drippings, and skim it off into a measuring cup.
3. Measure the amount of fat needed (according to the chart below) back into the roasting pan. (Meat juices should be used as part of liquid.)
4. Place over low heat.
5. Blend in flour and cook until bubbly, stirring constantly.
6. Brown fat and flour mixture slightly, if desired.
7. Remove pan from heat and add liquid gradually, stirring until smooth.
8. Return to heat and cook, stirring until mixture is thick; be sure to scrape brown particles from the bottom of the pan while cooking.
9. Simmer gently a few minutes, season to taste, and serve hot.

Fat Measurements for Gravy

	2 CUPS (473.2 mL) GRAVY (8 SERVINGS)	4 CUPS (0.9 L) GRAVY (16 SERVINGS)	6 CUPS (1.4 L) GRAVY (24 SERVINGS)
Fat	4 tablespoons (59.2 mL)	½ cup (118.3 mL)	¾ cup (177.5 mL)
Flour	4 tablespoons (59.2 mL)	½ cup (118.3 mL)	¾ cup (177.5 mL)
Liquid — broth, milk, or water	2 cups (473.2 mL)	4 cups (0.9 L)	6 cups (1.4 L)

Old-Fashioned Bread Stuffing

MAKES ENOUGH STUFFING TO FILL A 14- TO 18-POUND (6.4–8.2 KG), READY-TO-COOK TURKEY

4 cups (0.9 L) diced celery
1 cup (236.6 mL) finely chopped onion
1 cup (236.6 mL) butter
4 quarts (3.8 L) bread cubes, firmly packed (bread, 2 to 4 days old)
1 tablespoon (14.8 mL) salt
2 teaspoons (9.9 mL) poultry seasoning
½ teaspoon (2.5 mL) pepper
1½–2 cups (354.9–473.2 mL) broth, milk, or water

1. Cook celery and onion in butter over low heat, stirring occasionally, until onion is tender but not brown.
2. Meanwhile, blend bread cubes and salt, pepper, and seasoning. Add celery, onion, and butter; toss lightly to blend.
3. Pour the broth, milk, or water gradually over surface of bread mixture, tossing lightly. Add more seasoning, as desired. (To increase flavor, try reducing poultry seasoning and adding 1 teaspoon [4.9 mL) of sage, marjoram, and thyme and 1 tablespoon [14.8 mL] of parsley.)

   Note: Extra stuffing may be baked in a loaf pan or casserole for the last hour that the turkey is cooking. If desired, baste with pan drippings.

Corn Bread Stuffing

MAKES ENOUGH STUFFING FOR AN
18- TO 20-POUND (8.2–9.1 KG) TURKEY

1 pan corn bread, crumbled
1 package dry herbed stuffing mix (or bread crumbs and herbs) for 7-pound (3.2 kg) bird
2 cups (473.2 mL) chopped celery
1 cup (236.6 mL) chopped onion
½ cup (118.3 mL) butter
1 cup (236.6 mL) chicken stock
2 eggs, beaten

1. Combine corn bread and stuffing mix.
2. Cook celery and onion in butter until tender but not brown. Add to bread mixture.
3. Add stock and eggs, tossing lightly to blend well. (Add more stock for more moister dressing, if desired.)
4. Stuff lightly into neck and body cavity of turkey; truss.
5. Roast according to standard directions. Extra stuffing may be baked in a 1½-quart (1.4 L) casserole, covered, at 325°F (162.8°C) for about 1 hour.

Cranberry-Sausage Dressing

MAKES 6 TO 8 SERVINGS

2 cups (473.2 mL) fresh cranberries
1 cup (236.6 mL) orange juice
⅓ cup (78.9 mL) sugar
1 package (8 oz.; 23.2 g) corn bread stuffing mix
1 pound (0.5 kg) fresh pork sausage
½ teaspoon (2.5 mL) baking powder
1 cup (236.6 mL) finely chopped celery
½ cup (118.3 mL) finely chopped onion
1 egg
2 tablespoons (29.6 mL) water

1. In a saucepan, combine cranberries, orange juice, and sugar.
2. Bring to a boil, stirring until sugar dissolves. Boil for 5 minutes; cool.
3. Place stuffing mix in large bowl.
4. Break sausage into small pieces over stuffing mix; sprinkle with baking powder. Add celery, onion, and egg beaten with water. Toss gently, but mix well.
5. Fold in cooled, cooked cranberries.
6. Spoon mixture into 2-quart (1.8 L) casserole. Cover tightly and bake at 325°F (162.8°C) for 40 minutes. Remove cover and continue baking 15 minutes longer. Alternatively, use to stuff an 8- to 10-pound (3.6–4.5 kg) turkey.

Carving a Turkey

Allow 15 to 30 minutes between roasting and carving. This gives the juices time to be absorbed.

Method 1 (Traditional Method)

1. **Remove drumstick and thigh.** To remove drumstick and thigh, press leg away from body. Joint connecting leg to the hip often snaps free or may be severed easily with the point of a knife. Cut dark meat completely from body by following body contour carefully with the knife. ▶

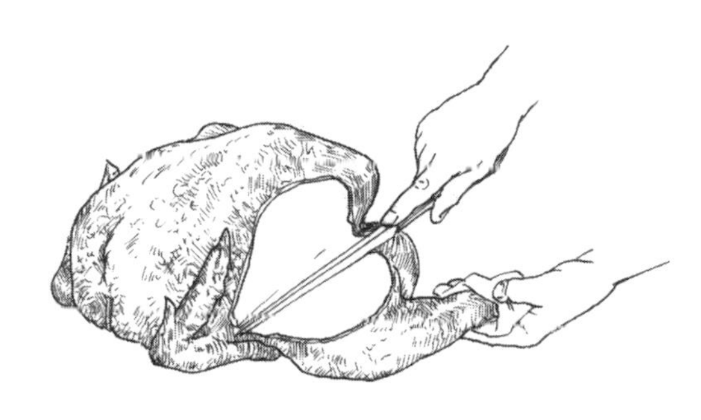

2. **Slice dark meat.** Place drumstick and thigh on cutting surface and cut through connecting joint. Both pieces may be individually sliced. Tilt drumstick to a convenient angle, slicing toward the table as shown. ▶

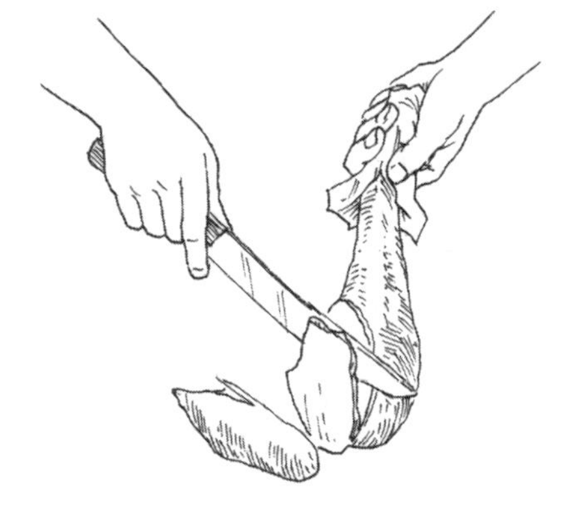

3. **Slice thigh.** To slice thigh meat, hold firmly on cutting surface with fork. Cut even slices parallel to the bone. ▶

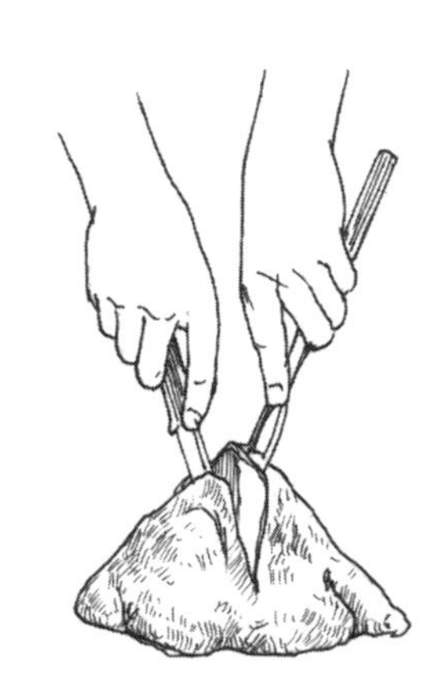

4. **Prepare breast.** In preparing breast for easy slicing, place knife parallel and as close to wing as possible. Make deep cut into breast, cutting right to the bone. This is your base cut. All breast slices stop at this vertical cut. ▶

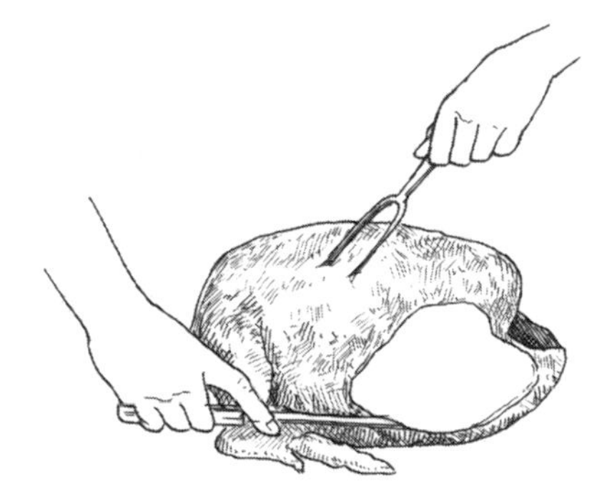

5. **Carve breast.** After making the base cut, carve downward, ending at the base cut. Start each new slice slightly higher up on breast. Keep slices thin and even. ▶

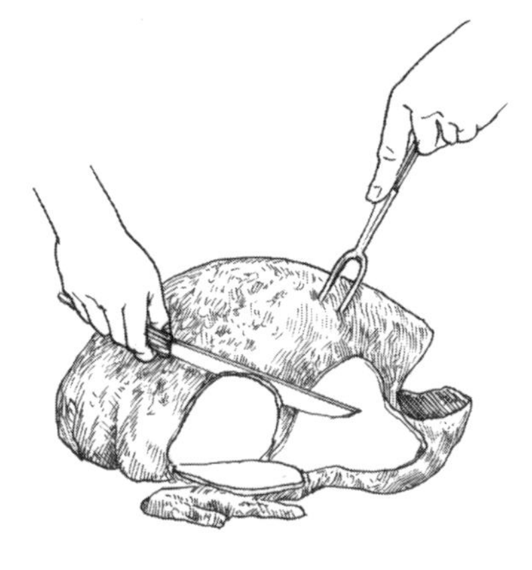

Method 2 (Kitchen-Carving Method)

1. **Remove drumstick and thigh by pressing leg away from body.** Joint connecting leg to backbone often snaps free or may be severed easily with the point of a knife. Cut dark meat completely from the body by following the body contour carefully with a knife. ▶

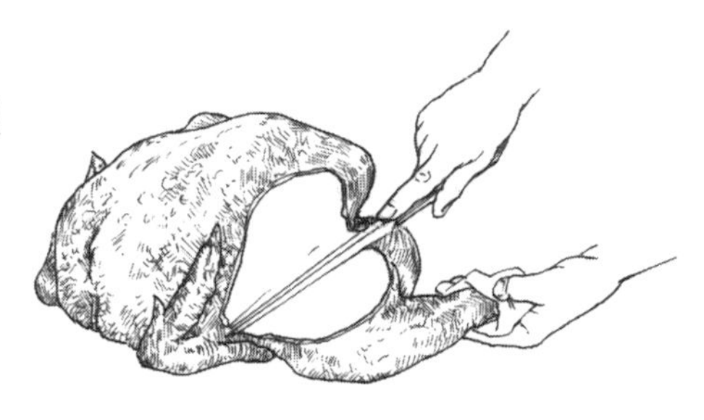

2. Place drumsticks and thigh on separate plate and **cut through connecting joint.** Both pieces may be individually sliced. Tilt drumstick to a convenient angle and slice toward the plate. ▶

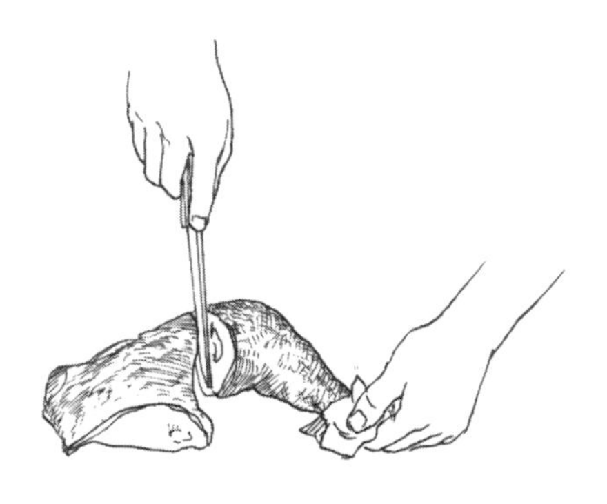

3. To slice thigh meat, hold firmly on plate with fork. **Cut even slices parallel to the bone.** ▶

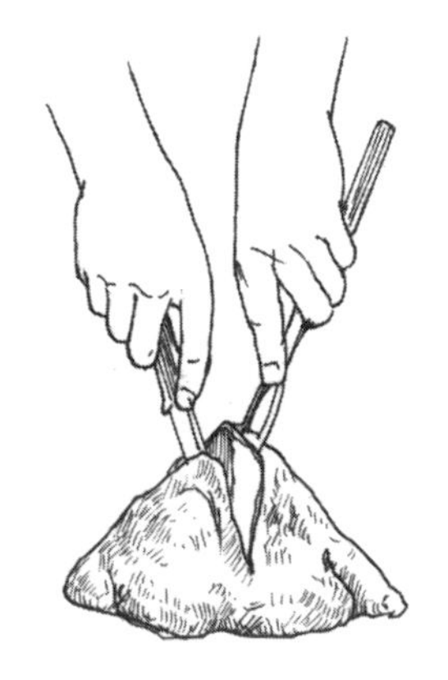

4. **Remove half of the breast** at a time by cutting along keel bone and rib cage with sharp knife. ▶

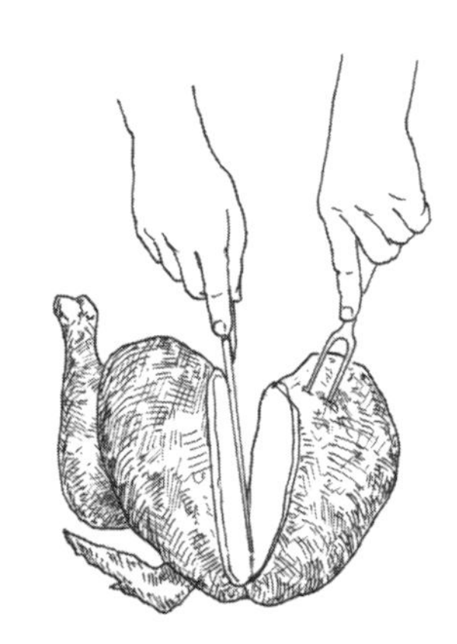

5. Place half breast on cutting surface and **slice evenly against the grain of the meat.** Repeat with second half breast when additional slices are needed. ▶

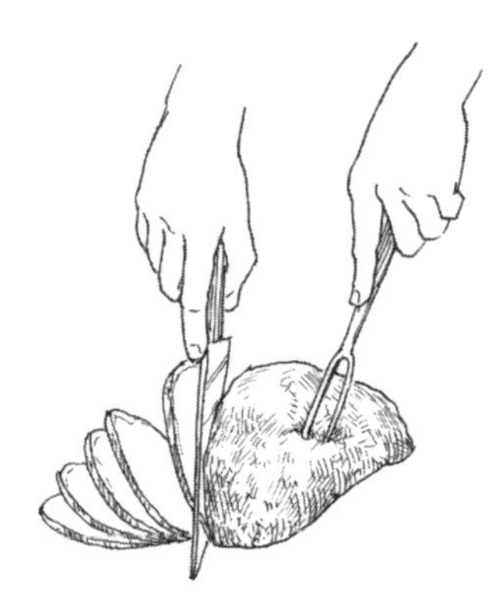

Serving Information

READY-TO-COOK WEIGHT	NUMBER OF SERVINGS
4–8 pounds (1.8–3.6 kg)	4–10
8–12 pounds (3.6–5.4 kg)	10–20
12–16 pounds (5.4–7.3 kg)	20–30
16–20 pounds (7.3–9.1 kg)	30–40
20–24 pounds (9.1–10.9 kg)	40–50

Uses for Turkey Parts

On occasion, birds may be injured or have leg problems or other conditions that make it advisable to cull them from the flock and salvage usable parts.

In addition, when dressing birds, you may find skin lesions, severe bruises, large breast blisters, or other blemishes that detract from the appearance of the dressed carcass. If the birds are otherwise healthy, they can be cut up into parts and used in many ways, depending on the age of the birds and your preference.

Possibilities include barbecuing, pan-frying, oven-frying, and deep-frying. In addition, turkey can be ground to make turkeyburgers.

Barbecuing

Turkeys are excellent for barbecuing. Turkey parts are barbecued in the same way as chicken. Charcoal fires and gas grills are excellent ways to barbecue a turkey; a gas-fired charcoal oven may work equally well. Turn the turkey frequently. With thick tomato-based or sweet-and-sour sauces, baste toward the end of the cooking time. Basting too soon with a thick sauce results in burned sauce. For light sauces, like the delicious one given on page 160, baste the turkey each time it is turned, from the beginning to the end of the cooking period. When applied in this manner, our barbecue sauce recipe gives the turkey an excellent flavor and nice light brown color. Cook the turkey until the meat thermometer reaches 175° to 180°F (79.4°–82.2°C).

Barbecue Sauce

Use your favorite sauce. If you don't already have a favorite, try making one from butter or margarine, water, salt, and vinegar. To cook approximately 25 pounds (11.3 kg) of turkey requires the following quantities of ingredients:

1 pound (0.5 kg) butter or margarine
1 quart (0.9 L) water
1 quart (0.9 L) cider vinegar
4 tablespoons (59.1 mL) salt

Combine the ingredients and heat until boiling. When the sauce boils, it's ready to use. For best results, apply the sauce to the turkey with a plastic sprinkling device. You might also apply it with a rag tied to the end of a stick. To serve the turkey as you would chopped pork or beef barbecue, remove the meat and skin from the turkey, chop it up, and add more sauce.

Note: Be sure to use sterile technique when working with raw meats of any kind. Reserve a separate container of sauce for use on meat once it's cooked, and use different utensils, too. Uncooked meat may contain salmonella or other harmful bacteria.

Pan-Frying

Birds from 4 to 9 pounds (1.8–4.1 kg) can be pan-fried successfully. Cut the turkey into pieces to yield the following parts:

- 2 drumsticks
- 2 thighs
- 4 breast pieces
- 2 wings
- 3 back pieces

How to Pan-Fry a Turkey

Oil or fat
2 tablespoons (29.6 mL) water

1. Heat ½ inch (1.3 cm) of oil or fat in a heavy skillet until a drop of water sizzles when it hits the pan.
2. Brown the meaty pieces first, then slip the less meaty pieces in between.
3. Turn as necessary to brown and cook evenly (about 20 minutes).
4. When pieces are browned nicely, reduce heat, add 2 tablespoons (29.6 mL), water, and cover tightly.
5. Cook slowly for 45 to 60 minutes, or until the thickest pieces are fork tender.
6. Turn pieces several times for even cooking and browning.
7. Uncover the pan the last 10 minutes for the skin to become crisp. Total cooking time is 1 to 1¼ hours.

Coating for Pan-Fried Turkey

¾ cup (177.5 mL) flour
1 teaspoon (4.9 mL) paprika
½ teaspoon (2.5 mL) oregano
2 teaspoons (9.9 mL) salt
¼ teaspoon (1.3 mL) pepper

1. Blend the ingredients in a bag for each 5 pounds (2.3 kg) of cut-up turkey. (For example, if you have a 15-pound (6.8 kg) turkey, triple the ingredient amounts, then blend.)
2. To coat evenly with the flour mixture, shake two or three pieces of turkey in the bag at a time. Remove the turkey from the bag. Reserve any leftover flour for use in gravy.

Oven-Frying

Smaller birds can be oven-fried.

1. Cut the turkey into parts and coat as in pan-frying. (See Coating for Pan-Fried Turkey, above.)
2. Use a shallow baking pan in a 350°F (176.7°C) oven.
3. Melt 1 cup of butter or margarine for each 5 pounds (2.3 kg) of turkey.
4. Place coated pieces in the pan, turning to coat all sides, then leave the skin side down. The turkey should fill the pan one layer deep without crowding or leaving any area of the pan exposed.
5. Bake for 45 minutes. Turn the pieces skin side up and continue baking for another 45 minutes, or until the meat is fork tender. Total cooking time is about 1½ hours.

Deep-Frying

According to the National Turkey Federation, deep-fried turkey is gaining in popularity. Special equipment is needed for frying turkey, and fortunately there are a number of companies that manufacture turkey fryers. Many companies that sell grilling equipment also sell turkey fryers. The equipment consists of the following:

- A 40- to 60-quart (36–54 L) pot
- Basket
- Burner
- Propane gas tank
- Candy thermometer to measure the oil temperature
- Meat thermometer to determine when the turkey is done

As with smoke-cooking, different marinades and seasonings with or without injection can be used. Always fry the turkey outside, never indoors. Also, do not set up the fryer on a wood deck or other area where it could cause a fire or other safety hazard. Smaller turkeys, 8 to 10 pounds (3.6–4.5 kg), and turkey parts, such as the breast, are best for frying.

1. You'll need approximately 5 gallons (9.5 L) of oil (more for larger birds). To determine the correct amount of oil, place the turkey in the basket and pot and cover with 1 to 2 inches (2.5–5.1 cm) of water. Remove the turkey and note the water level. This is the level to fill with oil. Discard water, and dry the pot well.
2. Fill pot to the correct level with oil. Heat the oil to 350°F (176.7°C). This takes 45 minutes to 1 hour.
3. While the oil is heating, prepare the turkey. Do not stuff a turkey that is going to be deep-fried.
4. When the oil is hot enough, place the turkey in the basket and lower it into the pot slowly. A whole turkey requires 3 minutes per pound (0.5 kg) to cook.
5. Remove the turkey and check the meat thermometer to see if it is done. The breast should be 170°F (76.7°C) and the thigh 180°F (82.2°C). Turkey parts require 4 to 5 minutes per pound (0.5 kg) to cook.

For additional information, contact the National Turkey Federation (see the references section in the appendix).

Turkeyburgers

Turkey can also be ground and used to make turkeyburgers. Try this interesting recipe.

4–5 PATTIES

1 pound (0.5 kg) ground turkey
8 soda crackers (crushed)
2 tablespoons (29.6 mL) ketchup
1 tablespoon (14.8 mL) lemon juice
1 tablespoon (14.8 mL) onion flakes
1 teaspoon (4.9 mL) Worcestershire sauce
½ teaspoon (2.5 mL) paprika
4 slices bacon

1. Combine the ground turkey with the other ingredients and shape into patties.
2. Wrap the bacon slices around the outer edge and secure with toothpicks.
3. Broil for 5 or 6 minutes on each side, or until burgers are completely done in the center.

The author thanks the National Turkey Federation for supplying information for this chapter. Visit its Web site for more turkey-cooking information: www.eatturkey.com.

References

Aho, W.A., and Talmadge, D.W. *Incubation and Embryology of the Chick*. Storrs, CT: Cooperative Extension Service, College of Agriculture and Natural Resources, University of Connecticut.

Aiello, S.E., et al. (eds). *The Merck Veterinary Manual*, 8th ed. Rahway, NJ: Merck & Co., Inc., 1998.

Department of Animal and Veterinary Sciences. *Clemson University Feathered Facts Series*. Clemson, SC: Clemson University.

Extension Poultrymen in New England. *Poultry Management and Business Analysis Manual for the 80s*. The New England Cooperative Extension Services, 1980.

Grimes, J.L., and Siopes, T.D. A survey of lighting practices in the U.S. turkey breeder industry. *Journal of Applied Poultry Research*. 8: 493–498, 1999.

Jordan, H.C., and Schwartz, L.D. *Home Processing of Poultry*. University Park, PA: The Pennsylvania State University, College of Agriculture, Extension Service.

Jordan, H.C. *Production of Market Turkeys*. Correspondence Courses in Agriculture, Course 106, Lesson 1, University Park, PA: The Pennsylvania State University Extension Service.

Marsden, S.J. *Turkey Production*. Agricultural Handbook No. 393. Washington, DC: Agricultural Research Service, USDA.

Mercia, L.S. *Killing, Picking and Cooking Poultry* (4-H publication). Burlington, VT: Vermont Extension Service, University of Vermont, 1976.

———. *The Small Turkey Flock*. Burlington, VT: The Vermont Extension Service, University of Vermont, 1977.

———. *Storey's Guide to Raising Poultry*. Pownal, VT: Storey Books, 2001.

Moyer, D.D. *Virginia Turkey Management*. Publication 302. Blacksburg, VA: Extension Division, Virginia Polytechnic Institute.

National Resource Council. *Nutrient Requirements of Poultry*, 9th ed. Washington, DC: National Academy Press, 1994.

National Turkey Federation, 1225 New York Avenue NW, Suite 400, Washington, DC 20005. Phone: 202-898-0100. Fax: 202-898-0203. *www.turkeyfed.org*

Schwartz, L.D. *Poultry Health Handbook*, 3rd ed. University Park, PA: The Pennsylvania State University, 1988.

Turkey Care Practices. California Poultry Workgroup, University of California, Cooperative Extension Service.

Diagnostic Laboratories by State

** Designates Laboratories Accredited by the American Association of Veterinary Laboratory Diagnosticians

Not all laboratories are full-service avian veterinary laboratories.

Alabama

State Veterinary Diagnostic Laboratory
Wire Road
P.O. Box 2209
Auburn, AL 36831-2209
334-844-4987

State Veterinary Diagnostic Laboratory
501 Usury Avenue
Boaz, AL 35957
256-593-2995

State Veterinary Diagnostic Laboratory
P.O. Box 409
Hanceville, AL 35077
256-352-8036

Alaska

State Federal Laboratory
500 S. Alaska Street
Palmer, AK 99645
907-745-3236
www.state.ak.us/dec/deh

Arizona

**Arizona Veterinary Diagnostic Laboratory
University of Arizona
2831 N. Freeway
Tucson, AZ 85705
520-621-2356
www.microvet.arizona.edu

Arkansas

Arkansas Livestock and Poultry Commission Laboratory
3559 N. Thompson Street
Springdale, AR 72764
501-751-4869

Arkansas Poultry Federation
P.O. Box 828
Springdale, AR 72764
501-375-8131

California

California Animal Health and Food Safety Laboratory Systems
Fresno Branch
2789 S. Orange Avenue
Fresno, CA 93725
559-498-7740

San Bernardino Laboratory
105 W. Central Avenue
P.O. Box 5579
San Bernardino, CA 92412
909-383-4287

Turlock Laboratory
1550 N. Soderquest
Turlock, CA 95380
209-634-5837

Colorado

Colorado State University,
Branch Diagnostic Laboratory
27847 Road 21
Rocky Ford, CO 81067
719-254-6382

Connecticut

**University of Connecticut
Department of Pathobiology
61 N. Eagleville Road
Box U-89
Storrs, CT 06269
860-486-3736

Delaware

Poultry and Animal Health
2320 S. Dupont Highway
Dover, DE 19901
800-282-8685

University of Delaware
Lasher Laboratory
Poultry Diagnostic Laboratory
Route 6, Box 48
Georgetown, DE 19947
302-856-1997
www.rec.udel.edu

Department of Animal and Food Sciences
University of Delaware
531 S. College Avenue
Newark, DE 19717
302-831-2524

Florida

** Kissimmee Diagnostic Laboratory
Florida Department of Agriculture
2700 N. John Young Parkway
Kissimmee, FL 34741
407-846-5200

Live Oak Diagnostic Laboratory
912 Nobles Ferry Road
Live Oak, FL 32060
904-362-1216

Georgia

Georgia Poultry Laboratory
Veterinary Poultry Diagnostic Laboratory
180 McClure Street
P.O. Box 349
Canton, GA 30114
770-479-2901

Georgia Poultry Laboratory
410 N. Park Avenue
Dalton, GA 30720
706-278-7306

Georgia Poultry Laboratory
150 Tom Freyer Drive
Douglas, GA 31535
912-384-3719

Georgia Poultry Laboratory
4457 Oakwood Road
Oakwood, GA 30566
770-535-5996

Hawaii

Veterinary Laboratory Branch
Hawaii Department of Agriculture
99-941 Halawa Valley Street
Aiea, HI 96701
808-483-7100

Idaho

Division of Animal Industries
2230 Old Penitentiary Road
Boise, ID 83712
208-332-8570

Illinois

**Animal Disease Laboratory
9732 Shattuc Road
Centralia, IL 62801
618-532-6701

**Animal Disease Laboratory
Illinois Department of Agriculture
2100 S. Lake Storey Road
P.O. Box 2100X
Galesburg, IL 61402-2100
309-344-2451

**Veterinary Diagnostic Lab
College of Veterinary Medicine
University of Illinois
Urbana, IL 61802
217-333-1620
www.cvm.uiuc.edu

Indiana

Animal Disease Diagnostic Laboratory
11367 E. Purdue Farm Road
Dubois, IN 47527
812-678-3401

**Animal Disease Diagnostic Laboratory
1175 ADDL
Purdue University
West Lafayette, IN 47907
765- 494-7440
www.addl.purdue.edu

Iowa

USDA National Veterinary Services Laboratory
P.O. Box 844
Ames, IA 50010
515-239-8200
www.aphis.usda.gov/vs/nvsl/index.html

**Veterinary Diagnostic Laboratory
College of Veterinary Medicine
Iowa State University
Ames, IA 50011
515-294-1950

Kansas

**Veterinary Diagnostic Laboratory
College of Veterinary Medicine
Kansas State University
Manhattan, KS 66506
785-532-5650
www.vet.ksu.edu

Kentucky

** Murray State University
Breathitt Veterinary Center
715 North Drive
P.O. Box 2000
Hopkinsville, KY 42240
270-886-3959
www.murraystate.edu/cit/bvc

**Livestock Disease Diagnostic Center
University of Kentucky
1429 Newtown Pike
Lexington, KY 40511
859-253-0571
www.ca.uky.edu/lddc

Maine

Animal Disease Diagnostic Laboratory
University of Maine
332 Hitchner Hall
Orono, ME 04469
207-581-2775/2788

Maryland

Maryland Department of Agriculture
Animal Health Laboratory
8077 Greenmead Drive
College Park, MD 20740
301-935-6074

Maryland Department of Agriculture
Animal Health Laboratory
1840 Rosemont Avenue
Frederick, MD 21702
301-663-9568
www.mda.state.md.us

Maryland Department of Agriculture
Animal Health Lab
P.O. Box 376
Oakland, MD 21550
301-334-2185

Maryland Department of Agriculture
Animal Health Laboratory
Quantico Road
P.O. Box 2599
Salisbury, MD 21801
410-543-6610

Michigan

**Animal Health Diagnostic Laboratory
Michigan State University
P.O. Box 30076
Lansing, MI 48909-1315
517-353-0635
www.ahdl.msu.edu

Minnesota

**Veterinary Diagnostic Laboratory
College of Veterinary Medicine
University of Minnesota
1333 Gortner Avenue
St. Paul, MN 55108
612-625-8787
www.mvdl.umn.edu

Poultry Testing Laboratory
622 Business Highway 71 NE
P.O. Box 126
Willmar, MN 56201
320-231-5170

Mississippi

Central Laboratory
P.O. Box 1510
Forest, MS 39074
601-469-4421

Mississippi Veterinary Diagnostic Laboratory
P.O. Box 4389
Jackson, MS 39296
601-354-6089

Missouri

Northwest Missouri Veterinary Diagnostic Laboratory
302 West Grand
Cameron, MO 64429
816-632-6595

**Veterinary Medical Diagnostic Laboratory
University of Missouri
P.O. Box 6023
Columbia, MO 65205
573-882-6811
www.cvm.missouri.edu/vmdl

State-Federal Cooperative
Animal Health Laboratory
216 El Mercado Plaza
Jefferson City, MO 65109
573-751-3460

Springfield Veterinary Diagnostic Laboratory
701 N. Miller Avenue
Springfield, MO 65802
417-895-6863

Montana

**Veterinary Diagnostic Laboratory
Department of Livestock
Diagnostic Laboratory Division
P.O. Box 997
Bozeman, MT 59771
406-994-4885

Nebraska

**Lincoln Diagnostic Laboratories
University of Nebraska–Lincoln
Fair Street and East Campus Loop
Lincoln, NE 68583
402-472-1434

Nevada

Nevada Department of Agriculture
Animal Disease Laboratory
350 Capitol Hill Avenue
Reno, NV 89502
775-688-1182, ext. 232

New Hampshire

New Hampshire Veterinary Diagnostic Lab
University of New Hampshire
Kendall Hall
129 Main Street
Durham, NH 03824
603-862-2726

New Jersey

New Jersey Animal Health Diagnostic Lab
John Fitch Plaza
P.O. Box 330
Trenton, NJ 08625
609-292-3965

New Mexico

New Mexico Veterinary Diagnostic Services
700 Camino de Salud NE
Albuquerque, NM 87106
505-841-2576

New York

Cornell Duck Research Laboratories
Veterinary Diagnostic Lab
192 Old Country Road
Eastport, NY 11941
631-325-0600

**Veterinary Diagnostic Laboratory
New York State College of Veterinary Medicine
Cornell University
Ithaca, NY 14853
607-253-3900

North Carolina

Animal Disease Diagnostic Laboratory
Paradise Road
P.O. Box 38
Edenton, NC 27932
252-482-3146

Hoyle C. Griffin Animal Disease Diagnostic Laboratory
401 Quarry Road
P.O. Box 2183
Monroe, NC 28111
704-289-6448
www.agr.state.nc.us/vet

**Rollins Animal Disease Diagnostic Laboratory
2101 Blue Ridge Road
Raleigh, NC 27607
919-733-3986
www.ncagr.com/vet/lab.htm

Rose Hill Animal Disease Diagnostic Laboratory
329 Yellowcut Road
P.O. Box 37
Rose Hill, NC 28458
910-289-2635

North Dakota

**Veterinary Diagnostic Laboratory
North Dakota State University
Van Es Hall
Fargo, ND 58105
701-231-8307
www.ndsu.nodak.edu/ndsu/veterinary_science/vetdiag

Ohio

**Animal Disease Diagnostic Laboratory
8995 E. Main Street
Reynoldsburg, OH 43068
614-728-6220
www.state.oh.us/agr/addl

Oklahoma

State Department of Health Laboratory
P.O. Box 24106
Oklahoma City, OK 73124
405-271-5070

**Animal Disease Diagnostic Laboratory
College of Veterinary Medicine
Oklahoma State University
Stillwater, OK 74078
405-744-6623
www.cvm.okstate.edu

Oregon

OSU Veterinary Diagnostic Laboratory
College of Veterinary Medicine
Oregon State University
P.O. Box 429
Corvallis, OR 97339
541-737-3261
www.vet.orst.edu

Oregon State Department of Agriculture
Animal Health Laboratory
635 Capital Street NE
Salem, OR 97301
503-986-4686
www.oda.state.or.us

Pennsylvania

**The Pennsylvania Veterinary Laboratory (PADLS)
2305 N. Cameron Street
Harrisburg, PA 17110
717-772-2852

Diagnostic Laboratory
New Bolton Center
382 West Street Road
Kennett Square, PA 19348
610-444-5800

Animal Diagnostic Laboratory
Orchard Road
The Pennsylvania State University
University Park, PA 16802-1110
814-863-0837

South Dakota

Department of Veterinary Science
Animal Disease Research and Diagnostic Laboratory
South Dakota State University
Box 2175
Brookings, SD 57007
605-688-5171
www.vetsci.sdstate.edu

Tennessee

University of Tennessee College of Veterinary Medicine
Diagnostic Services
P.O. Box 1071
Knoxville, TN 37901
865-974-7262
www.vet.utk.edu

C.E. Kord Animal Disease Laboratory
Melrose Station
P.O. Box 40627
Nashville, TN 37204
615-837-5125
www.state.tn.us/agriculture

Texas

**Texas Veterinary Medical Diagnostic Laboratory
Texas A&M University
Drawer 3040
College Station, TX 77841
409-845-9000

Utah

Utah State University
Veterinary Diagnostic Laboratory
North Logan, UT 84322
435-797-1900

USU Provo Veterinary Laboratory
2031 S. State Street
Provo, UT 84606
801-373-6383

Virginia

Harrisonburg Laboratory
116 Reservoir Street
Harrisonburg, VA 22801
540-434-3897
www.vdacs.state.va.us

Ivor Laboratory
34591 General Mahone Boulevard
Ivor, VA 23866
757-859-6221
www.vdacs.state.va.us

Lynchburg Laboratory
4832 Tyreeanna Road
Lynchburg, VA 24504
804-947-2518
www.vdacs.state.va.us

Warrenton Laboratory
272 Academy Hill Road
Warrenton, VA 20186
540-347-6385
www.vdacs.state.va.us

Wytheville Laboratory
250 Cassel Road
Wytheville, VA 24382
540-228-5501
www.vdacs.state.va.us

Washington

**Animal Disease Diagnostic Laboratory
Washington State University
College Station
P.O. Box 2037
Pullman, WA 99165
509-335-9696
www.vetmed.wsu.edu

Avian Health Laboratory
Washington State University
Puyallup, WA 98371
253-445-4537
www.vetmed.wsu.edu

West Virginia

State Federal Cooperative
Animal Health Laboratory
1900 Kanawha Blvd. East
Charleston, WV 25305-0172
304-558-2214

Wisconsin

Wisconsin Veterinary Diagnostic Laboratory
1521 E. Guy Avenue
Barron, WI 54812
715-637-3151

**Wisconsin Veterinary
Diagnostic Laboratory
University of Wisconsin
6101 Mineral Point Road
Madison, WI 54705
608-262-5432

Wyoming

**Wyoming State Veterinary Laboratory
1174 Snowy Range Road
Laramie, WY 82070
307-742-6638

Canada

**Animal Health Monitoring Lab
1767 Angus Campbell Road
Abbotsford, BC V3G 2M3
CANADA
604-556-3135

**University of Guelph
Animal Health Laboratory
P.O. Box 3612
Guelph, ON N1H 6R8
CANADA
519-824-4120

Diagnostic Services
Atlantic Veterinary College
University of Prince Edward Island
550 University Avenue
Charlottetown, PE C1E 1Z4
CANADA
902-566-0863

Cooperative Extension System Offices by State

The Cooperative Extension System is an excellent source of information on poultry and other agricultural enterprises. Extension personnel are located in county or area offices in each state, as well as at agricultural colleges of the state universities. Poultry specialists are located at many state universities, but not all of them. In some large poultry-producing states, specialists may be located in various areas of the state, as well as on state university campuses. For information, contact the Cooperative Extension System office nearest you.

Alabama

Auburn University
Poultry Science Department
236 Animal Science Building
Auburn, AL 36849-5416
334-844-2600
www.ag.auburn.edu/dept/ph/index.html
www.aces.edu

Tuskegee University
Room 202 Morrison/Mayberry Hall
Tuskegee, AL 36088
334-724-4441

Alaska

Cooperative Extension Service
College of Rural Alaska
University of Alaska Fairbanks
P.O. Box 756180
Fairbanks, AK 99775-6180
907-474-7246
www.uaf.edu/coop-ext

Arizona

University of Arizona
Forbes Building, Room 301
Tucson, AZ 85721
520-621-7209
www.ag.arizona.edu/extension

Arkansas

University of Arkansas
Poultry Science Department
Fayetteville, AR 72701
501-575-4952
www.uark.edu/depts/posc
www.uaex.edu

University of Arkansas at Pine Bluff
1890 Cooperative Extension Service
1200 N. University Drive
Pine Bluff, AR 71601
870-543-8529/8534
www.uapb.edu

California

University of California at Davis
Department of Avian Science
Davis, CA 95616-8521
530-752-3519
animalscience.ucdavis.edu/extension/avian/

University of California
111 Franklin Street, 6th Floor
Oakland, CA 94607-5200
510-987-0060
danr.ucop.edu

Colorado

Colorado State University
1 Administration Building
Fort Collins, CO 80523-4040
970-491-6281
www.colostate.edu/Depts/CoopExt/

Connecticut

University of Connecticut
Young Building, Room 231
1376 Storrs Road, U-36
Storrs, CT 06269
860-486-1987
www.canr.uconn.edu/ces/index.html

Delaware

University of Delaware
113 Townsend Hall
Newark, DE 19717
302-831-2501
ag.udel.edu

Florida

University of Florida
Department of Dairy and Poultry Sciences
P.O. Box 110920
Gainesville, FL 32610
352-392-1981
dps.ufl.edu/

Georgia

University of Georgia
Department of Poultry Science
4 Towers Building
Athens, GA 30602
706-542-1325
www.uga.edu/~poultry/
www.uga.edu/caes/

Hawaii

University of Hawaii
3050 Maile Way, Room 202
Honolulu, HI 96822
808-956-8234
www2.ctahr.hawaii.edu

Idaho

University of Idaho
P.O. Box 442332
Moscow, ID 83844-2332
208-885-6639

Illinois

University of Illinois
214 Mumford Hall
1301 W. Gregory Drive
Urbana, IL 61801
217-333-5900
www.extension.uiuc.edu

Indiana

Purdue University
Department of Animal Science
1151 Smith Hall
West Lafayette, IN 47907-1151
765-494-8011
ag.ansc.purdue.edu/poultry/

Iowa

Iowa State University
218 Beardshear Hall
Ames, IA 50011
515-294-6192
www.exnet.iastate.edu

Kansas

Kansas State University
123 Umberger Hall
Manhattan, KS 66506
785-532-5820
www.oznet.ksu.edu

Kentucky

University of Kentucky
Department of Animal Science
604 W.P. Garrigus Building
Lexington, KY 40546
859-257-4302
www.uky.edu/Agriculture/AnimalSciences/Poultry/ukpoultry.htm
www.ca.uky.edu

Louisiana

Louisiana State University
Department of Poultry Science
Ingram Hall
Baton Rouge, LA 70894
504-388-4406
www.coa.lsu.edu
www.lsuagcenter.com

Maine

University of Maine
5741 Libby Hall, Room 102
Orono, ME 04469
207-581-2811
www.umext.maine.edu

Maryland

University of Maryland at College Park
Department of Animal and Avian Sciences
College Park, MD 20742
301-405-1373
ansc.umd.edu/ansc2000/index.html

University of Maryland, Eastern Shore
Backbone Road
Princess Anne, MD 21853
410-651-6206

Massachusetts

University of Massachusetts
Extension Office
Draper Hall
University of Massachusetts
Amherst, MA 01003
413-545-4800
www.umass.edu/umext

Michigan

Michigan State University
108 Agriculture Hall
East Lansing, MI 48824
517-355-2308

Minnesota

University of Minnesota
Coffey Hall, Room 240
1420 Eckles Avenue
St. Paul, MN 55108
612-624-2703

Mississippi

Alcorn State University
1000 ASU Drive, 479
Lorman, MS 39096
601-877-6128
www.alcorn.edu

Mississippi State University
Department of Poultry Science
P.O. Box 5188
Mississippi State, MS 39762
662-325-3416
www.msstate.edu/dept/poultry/
www.ext.msstate.edu

Missouri

University of Missouri
309 University Hall
Columbia, MO 65211
573-882-7754
outreach.missouri.edu/

Montana

Montana State University
P.O. Box 172040
115 Culbertson Hall
Bozeman, MT 59717-2040
406-994-6647
www.montana.edu

Nebraska

University of Nebraska
211 Agriculture Hall
Lincoln, NE 68583
402-472-2966
ianrhome.unl.edu/extension1.html

Nevada

University of Nevada at Reno
Cooperative Extension
National Judicial College 118, Mail Stop 404
Reno, NV 89557-0106
775-784-7070
www.nce.unr.edu

New Hampshire

University of New Hampshire
Cooperative Extension
Taylor Hall
59 College Road
Durham, NH 03824
603-862-1520
ceinfo.unh.edu

New Jersey

Rutgers Cooperative Extension
Rutgers, The State University of New Jersey
88 Lippman Drive
New Brunswick, NJ 08901
732-932-5000
www.rce.rutgers.edu

New Mexico

New Mexico State University
Department 3AE
P.O. Box 30003
Las Cruces, NM 88003
505-646-3016
cahe.nmsu.edu

New York

Cornell University
365 Roberts Hall
Ithaca, NY 14853
607-255-2116
www.cce.cornell.edu

North Carolina

North Carolina A&T State University
P.O. Box 21928
Greensboro, NC 27420
336-334-7691
www.ag.ncat.edu

North Carolina State University
Extension Poultry Science
P.O. Box 7603
Raleigh, NC 27695-7608
919-515-2621
www.ces.ncsu.edu/depts/poulsci

North Dakota

North Dakota State University
315 Morrill Hall
P.O. Box 5437
Fargo, ND 58105
701-231-8944
www.ext.nodak.edu

Ohio

Ohio State University Extension
Ohio State University
3 Agricultural Administration Building
2120 Fyffe Road, Room 4
Columbus, OH 43210
614-292-4067
ohioline.ag.ohio-state.edu

Oklahoma

Oklahoma State University
Department of Animal Science
101 Animal Science Building
Stillwater, OK 74078-6051
405-744-5398
www.ansi.okstate.edu/
www.dasnr.okstate.edu

Oregon

Oregon State University
Department of Animal Sciences
101 Ballard Extension Hall
Covallis, OR 97331-6702
541-737-5066
www.orst.edu/Dept/animal-sciences/poultext.htm
osu.orst.edu/extension/

Pennsylvania

Pennsylvania State University
Department of Poultry Science
213 William N. Henning Building
University Park, PA 16802
814-863-3411
poultry.cas.psu.edu/

Rhode Island

University of Rhode Island
9 E. Alumni Avenue, Room 137
Kingston, RI 02881
401-874-2970
www.uri.edu/ce/

South Carolina

Clemson University
Animal and Veterinary Sciences Department
129 Poole Agricultural Center
Box 340361
Clemson, SC 29634-0361
864-656-3427
cufp.clemson.edu/avs/
virtual.clemson.edu/groups/public

South Dakota

South Dakota State University
Agricultural Hall 154
P.O. Box 2207D
Brookings, SD 57007
605-688-4792

Tennessee
The University of Tennessee
121 Morgan Hall
P.O. Box 1071
Knoxville, TN 37901
865-974-7114
www.utextension.utk.edu

Texas
Texas A&M University
Poultry Science Department
Kleberg Center, Room 101
College Station, TX 77843-2472
979-845-1931
gallus.tamu.edu/departmental.html

Utah
Utah State University
4900 Old Main Hill
Logan, UT 84322
435-797-2200
www.ext.usu.edu

Vermont
University of Vermont
College of Agriculture and Life Sciences
601 Main Street
Burlington, VT 05405
802-656-2990
ctr.uvm.edu/ext/

Virginia
Virginia Polytechnic Institute and State University
Department of Animal and Poultry Sciences
101 Hutcheson Hall
Blacksburg, VA 24061-0306
540-231-9185
www.apsc.vt.edu
www.ext.vt.edu

Washington
Washington State University
421 Hulbert Hall
P.O. Box 646230
Pullman, WA 99164
509-335-4561
cahe.wsu.edu

West Virginia
West Virginia University
P.O. Box 6031
Morgantown, WV 26506-6031
304-293-5691
www.wvu.edu\~exten

Wisconsin
University of Wisconsin
Department of Animal Sciences
Animal Science Building
1675 Observatory Drive
Madison, WI 83706-1284
608-263-4300
www.poultry.wisc.edu
www.uwex.edu/ces/cty/grant/index.html

Wyoming
University of Wyoming
P.O. Box 3354
Laramie, WY 82071
307-766-5124

Associations

Alabama Poultry & Egg Association
P.O. Box 240
Montgomery, AL 36101
334 -265-2732
www.alabamapoultry.org

Alberta Turkey Producers
212 8711A-50 Street
Edmonton, AB T6B 1E7
CANADA
780-465-5755
www.abturkey.ab.ca

American Association of Avian Pathologists
382 West Street Road
Kennett Square, PA 19348
610-444-4282
www.vm.iastate.edu/aaap

American Association of Meat Processors
P.O. Box 269
Elizabethtown, PA 17022
717-367-1168
www.aamp.com

American Farm Bureau Federation
225 Touhy Avenue
Park Ridge, IL 60068
847-685-8600
www.fb.com

American Society of Agricultural Engineers
2950 Niles Road
St. Joseph, MI 49085-9659
616-429-0300
www.asae.org

American Veterinary Medical Association
1931 N. Meacham Road, Suite 100
Schaumburg, IL 60173
847-925-8070
www.avma.org

Animal Health Institute
1325 G Street NW, Suite 700
Washington, DC 20005
202-637-2440
www.ahi.org

Arkansas Poultry Federation
P.O. Box 1446
Little Rock, AR 72203
501-375-8131

California Poultry Industry Federation
3117A McHenry Avenue
Modesto, CA 95350
888-822-4004
www.cpif.org

Delmarva Poultry Industry, Inc.
RD 6, Box 47
Georgetown, DE 19947
302-856-9037
www.dpichicken.org

Florida Poultry Federation
4508 Oak Fair Boulevard, Suite 290
Tampa, FL 33610
813-628-4551

Food Processing Machinery & Supplies Association
200 Daingerfield Road
Alexandria, VA 22314
800-331-8816
www.fpmsa.org

Food Safety Consortium
110 Agriculture Building
University of Arkansas
Fayetteville, AR 72701
501-575-5647

Georgia Poultry Federation
P.O. Box 763
Gainesville, GA 30503
770-532-0473

Indiana State Poultry Assn.
Purdue University
1151 Lilly Hall, Room G117
West Lafayette, IN 47906
765-494-8517

Kansas Poultry Association
Kansas State University
Dept. of Animal Sciences
139 Call Hall
Manhattan, KS 66506-1600
785-532-1201
www.oznet.ksu.edu/

Kentucky Poultry Federation
P.O. Box 21829
Lexington, KY 40522
606-266-8375

Louisiana Poultry Federation
214 Knapp Hall
Louisiana State University
Baton Rouge, LA 70803
225-388-2219
www.agctr.lsu.edu

Meat & Poultry Association of Hawaii
311 Pacific Street
Honolulu, HI 96817
808-585-2900

Midwest Poultry Federation
2380 Wycliff Street
St. Paul, MN 55114
651-646-4553

Mississippi Poultry Association
P.O. Box 13309
Jackson, MS 39236
601-355-0248

Missouri Poultry Federation
225 E. Capitol Avenue
Jefferson City, MO 65101
573-761-5610

Food Distributors International
201 Park Washington Court
Falls Church, VA 22046
703-532-9400
www.fdi.org

National Association of Meat Processors
1920 Association Drive, Suite 400
Reston, VA 20191-1547
800-368-3043
www.namp.com

National Association of State Depts. of Agriculture
1015 15th Street NW, Suite 930
Washington, DC 20005
202-296-2622
www.eatchicken.com

National Grocers Association
1825 Samuel Morse Drive
Reston, VA 20190
703-437-5300
www.nationalgrocers. org

National Poultry & Food Distributors Assoc.
958 McEver Road Ext., Suite B5
Gainesville, GA 30506
770-535-9901
www.npfda.org

National Renderers Association
801 N. Fairfax Street, Suite 207
Alexandria, VA 22314
703-683-0155
www.renderers.org

National Restaurant Association
1200 17th Street NW
Washington, DC 20036
800-424-5156
www.restaurant.org

National Turkey Federation
1225 New York Avenue NW, Suite 400
Washington, DC 20005
202-898-0100
www.eatturkey.com

Nebraska Poultry Industry
A103 Animal Science
University of Nebraska Lincoln
P.O. Box 830908
Lincoln, NE 68583-0908
402-472-2051

North Carolina Poultry Federation
4020 Barrett Drive, Suite 102
Raleigh, NC 27609
919-783-8218

Pacific Egg & Poultry Association
1521 I Street
Sacramento, CA 95814
916-441-0801
www.pacificegg.org

Packaging Machinery Manufacturers Institute
4350 N. Fairfax Drive, Suite 600
Arlington, VA 22203
703-243-8555
www.packexpo.com

Poultry Industry Council
RR 2, 483 Arkell Road
Guelph, ON N1H 6H8
CANADA
519-837-0284
www.easynet.ca/~pic

Poultry Science Association
1111 N. Dunlap Avenue
Savoy, IL 61874
217-356-3182
www.psa.uiuc.edu

South Carolina Poultry Federation
1921A Pickens Street
Columbia, SC 29201
803-779-4700

South Dakota Poultry Industries Association
Animal and Range Sciences Department, SDSU
Box 2170
Brookings, SD 57007
605-688-5409
www.abs.sdstate.edu/ars/index.htm

Texas Poultry Federation
P.O. Box 9589
Austin, TX 78766
512-451-6816

U.S. Poultry & Egg Association
1530 Cooledge Road
Tucker, GA 30084
770-493-9401
www.poultryegg.org

United States Animal Health Association
P.O. Box K227
8100 Three Chopt Road
Richmond, VA 23288
804-285-3210
www.usaha.org

Virginia Poultry Federation
P.O. Box 552
Harrisonburg, VA 22801
540-433-2451

Sources of Supplies and Equipment

Supplies and equipment needed for the small poultry flock may be found at the local feed store, hatchery, or other agriculture-supply outlets. Some of the large mail-order houses, such as Sears, also offer agriculture supplies.

The following sources of stock, equipment, and veterinary supplies are listed for your convenience. No endorsement is expressed or implied.

Agri-Equipment International, Inc.
P.O. Box 8401
Greenville, SC 29604
877-550-4709
Bags for dressed poultry.

Ashley Machine, Inc.
901 N. Carver Street
P.O. Box 2
Greensburg, IN 47240
812-663-2180
Killing cones, knives, dressing equipment, processing equipment.

Beacon Systems Manufacturing Co.
Route 1, Box 354-A
Buffalo, MO 65622
417-345-2266
Brooders, egg baskets, nests, egg candlers (hand), scales, feeders, waterers.

Big Dutchman
P.O. Box 1017
Holland, MI 49423-1017
616-392-5981
Brooders, nests, feeders, waterers.

Brower Equipment Co.
P.O. Box 2000
Houghton, IA 52631-2000
319-469-4141
Brooders, egg candlers (hand), scales, washers, incubators (small), killing cones, knives, dressing equipment, processing equipment.

Cutler's Supply
3805 Washington Road
Carsonville, MI 48419
810-657-9450
General supplies.

Dussek Campbell, Inc.
National Wax Division
P.O. Box 549
Skokie, IL 60076
847-679-6300
www.dussekwax.com
Wax for defeathering.

First State Packaging, Inc.
511 Naylor Mill Road
Salisbury, MD 21801
410-546-1008
Bags for dressed poultry.

First State Veterinary Supply
P.O. Box 190
Parsonburg, MD 21849
800-950-8387

G.Q.F. Manufacturing Company
2343 Louisville Road
P.O. Box 1552
Savannah, GA 31402-1552
912-236-0651
www.gqfmfg.com
Wire pens, incubators (small), brooders, supplies.

Humidaire Incubator Company
217 W. Wayne Street
P.O. Box 9
New Madison, OH 45346
800-410-6925
www.411web.com (keyword "humidaire")
Redwood cabinet incubators, small incubators.

Inman Hatcheries
P.O. Box 616
Aberdeen, SD 57402
800-843-1962
www.inmanhatcheries.com
Chicks of various breeds.

Jeffers Vet Supply
P.O. Box 100
Dothan, AL 36302
800-533-3377
www.jefferspet.com
Medications and general supplies.

Kent Co., Inc.
3030 NE 188th Street
Miami, FL 33180
305-944-4041
Egg candlers (hand), scales, washers, incubators (small), processing equipment, smoke houses.

Lyon Electric Company
1690 Brandywine Avenue
Chula Vista, CA 91911-6021
619-216-3400
Incubators, brooders, parts, and accessories; cannibalism-control equipment.

Max-Flex
U.S. Route 219
Lindside, WV 24951
800-356-5458
www.maxflex.com
Electroplastic fencing.

McMurray Hatchery
191 Closz Drive
Webster City, IA 50595
515-832-3280
www.mcmurrayhatchery.com
Chicks, general supplies, books.

NASCO Farm & Ranch
901 Janesville Avenue
Fort Atkinson, WI 53538-0901
800-558-9595
www.enasco.com
Full line of general supplies; egg candlers (hand), scales, washers; killing cones, knives, and other dressing equipment.

Northco Industries, Inc.
P.O. Box 718
Luverne, MN 56156-0718
507-283-4411
Brooders.

Omaha Vaccine Company
P.O. Box 7228
Omaha, NE 68107
800-367-4444
www.omahavaccine.com
Medications and supplies.

Peterson Poultry Supplies
P.O. Box 39
Wallburg, NC 27373
336-769-0392
Full line of supplies and books.

Petersyme Incubator Co.
P.O. Box 308
Gettysburg, OH 45328-0308
888-255-0067
Incubators (small).

Pickwick Co.
1870 McCloud Place NE
Cedar Rapids, IA 52402
319-393-7443
Limited supplies.

Premier Fence Supply
2031 300th Street
Washington, IA 52353
800-282-6631
Electroplastic fencing.

Safeguard Products, Inc.
P.O. Box 8
New Holland, PA 17557
800-433-1819
www.safeguardproducts.com
Wire cages, components, tools, accessories.

Sand Hill Preservation Center
1878 230th Street
Calamus, IA 52729
319-246-2299
Chicks in rare and heirloom breeds.

Shenandoah Manufacturing Co., Inc.
1070 Virginia Avenue
Harrisonburg, VA 22802
800-476-7436
Brooders, nests, feeders, waterers.

Smith Poultry & Game Bird Supplies
14000 West 215th Street
Bucyrus, KS 66014
913-879-2587
www.poultrysupplies.com
Books and general supplies.

Strecker Supply Co.
P.O. Box 190
Parsonburg, MD 21849
800-765-0065
Medications.

Stromberg's Chicks and Gamebirds Unlimited
P.O. Box 400
Pine River, MN 56474
800-720-1134
www.strombergschickens.com
Chicks; full line of supplies and books; feeders, waterers, incubators (small).

Waterford Corporation
404 North Link Lane
Fort Collins, CO 80524
800-525-4952
www.waterfordcorp.com
Electroplastic fencing.

Glossary

Air cell Air space in the egg, usually in the large end.

Albumen The white of an egg consisting of thick and thin layers.

Alectors chukar Chukar partridge, a nonnative gamebird of the western United States, also a popular preserve bird.

Allantois Respiratory and excretory organ of bird embryos prior to lung development and activation.

Ambient temperature Actual outside temperature.

Amnion A membranous sac enclosing and protecting the embryo that holds the amniotic fluid.

Anas platyrhynchos Wild mallard, ancestor of today's domestic ducks.

Androgen A sex hormone produced in the testes and characterized by its ability to stimulate the development of sex characteristics in the male.

Anser anser and ***Anser cygnoides*** The gray lag goose of Europe and China, respectively.

Anterior The front part.

Antioxidant Compounds that reduce free radicals in the body; also, compounds used to prevent rancidity of fats or the destruction of fat-soluble vitamins.

Avian Of or pertaining to birds.

Bacteria Microscopic single-celled organisms.

Beak Upper and lower mandibles of chickens, peafowl, pheasants, turkeys, and so on.

Beak trimming Removal of the upper and/or lower tips of the beak to prevent cannibalism and improve feed efficiency.

Bits/Rings Attachments for mandible to prevent cannibalism.

Blastoderm The collective mass of cells of a fertilized ovum from which the embryo develops.

Blastodisc The germinal spot on the ovum from which the blastoderm develops after the ovum is fertilized by the sperm.

Blinders/Specks Attachments for upper mandible to partially block vision. Used to prevent cannibalism.

Breast The forward part of the body between the neck and keel bone.

Breast blister Swollen, discolored area or sore in the area of the keel bone.

Brooder Heat source for starting young birds.

Broodiness Tendency toward the maternal instinct that causes females to set or want to hatch eggs.

Bursa fabricious A glandular organ located dorsally to the cloaca, important to the immunology of the bird, regresses as the bird matures.

Candle To determine interior condition of the egg through the use of a special light in a dark room.

Cannibalism In the poultry industry, this term refers to the habit of one bird's picking another to the point of injury or death. Can occur as toe picking, feather picking or pulling, vent picking, head picking, or tail picking.

Caponization Surgical removal of the testes from a bird.

Carbohydrate A class or type of nutrient that serves as an energy source and is derived from plant sources, such as grain.

Caruncles The fleshy, nonfeathered area on the neck of a turkey.

Ceca Two blind pouches located at the junction of the lower small intestines and the rectum that aid the digestion of birds, especially when fed highly fibrous diets.

Chalazae Prolongations of the thick inner albumen that are twisted like ropes at both ends of the yolk; they anchor the yolk in the center of the eggshell cavity.

Chick Young chicken, quail, or pheasant.

Cholecalciferol Vitamin D_3 needed for the absorption and deposition of calcium.

Chorion A membrane enveloping the embryo, external to and enclosing the amnion.

Chromosomes A series of paired bodies in the nucleus of a cell that contain DNA and are responsible for hereditary characteristics; constant in any one kind (species) of plant or animal.

Cloaca In birds, the common chamber or receptacle for the digestive, urinary, and reproductive tracts.

Coccidiostat A drug used to control or prevent coccidiosis, a disease of poultry caused by protozoa.

Colinus virginianus The Bobwhite quail.

Columba livia domestica The domestic pigeon.

Comb Specialized structure on the top of the chicken's head.

Confinement-rearing Rearing of animals in an enclosed or semi-enclosed building, such as a barn or shed.

Coturnix coturnix A quail of Europe, Africa, and Asia; also known as the subspecies "Japanese" quail and "pharaoh" quail.

Crop An enlargement of the esophagus in which food is stored and prepared for digestion.

Crumble Form in which some feeds are supplied; it refers to animal feed that has been pelleted and then reground or crumbled into small bits.

Cull A bird not suitable to be kept as a breeder or market bird.

Culling The act of removing unsuitable birds from the flock.

Cuticle External waxy covering or coating of the egg.

Cygnet A young swan.

***Cygnus* species** Genus of swans.

Darkling beetle A black beetle that also exists as the lesser mealworm, the larval stage of the darkling beetle, which thrives in poultry bedding material such as pine shavings (litter).

Desnooding Removal of the fleshy appendage from a turkey's head, usually done at the hatchery or on the day of hatching.

Doral Of, on, or near the back.

Down Hairlike body feathers covering newly hatched poultry, including turkey poults.

Dry-bulb thermometer Expresses a temperature in number of degrees in Fahrenheit or Celsius.

Duckling A young duck.

Dubbing Removal of the comb from a chicken, usually done at the hatchery.

Egg The female reproductive cell (ovum) surrounded by a protective calcium shell and, if fertilized by the male reproductive cell (sperm) and properly incubated, capable of developing into a new individual of the species.

Egg tooth Temporary extension on the chick or poult's upper beak used to crack the shell at hatching

Embryo An organism in the early stages of development, as before hatching from the egg.

Esophagus The tubular structure leading from the mouth to the glandular stomach.

Evaporation Changing of moisture (liquid) into a vapor (gas).

Fertile Capable of reproducing.

Fertilization Penetration of the female sex cell (ovum) by the male sex cell (sperm) resulting in the fusion of the cell nuclei.

Flight feathers The large primary and secondary feathers of the wings.

Follicle A developing yolk on the ovary.

Foot-candle A measurement of the intensity of a light; technically, it is the amount of light striking every point on a segment of the inside of a sphere or on a surface area of 1 square foot all parts of which are 1 foot from an international candle (a candle of a specified size that emits a specified amount of light).

Gallus gallus Red jungle fowl, ancestor of the domestic chicken, still found in the South Pacific.

Gallus domesticus The domestic chicken.

Gamebird Breeds of fowl utilized on hunting preserves; usually refers to Bobwhite quail, Ringnecked pheasants, and chukars. Sometimes confused with "gaming" birds, such as fighting cocks and game cocks.

Gene Element in the chromosome that transmits hereditary characteristics.

Gizzard The muscular digestive organ in birds used for grinding the digesta, located between the proventriculus and intestine.

Gonad Gland that produces reproductive cells; the ovary or testis.

Gosling A young goose.

Hasp A hinged fastener for a door that is passed over a staple and secured by a pin, bolt, or padlock.

Hatching egg A fertilized egg intended for incubation.

Hatchability of fertile eggs Number of poults hatched compared with the number of fertile eggs set.

Hatchability of all eggs Number of poults hatched compared with the number of all eggs set.

Hen The female of most fowl, including turkeys.

Hock The joint where the shank (metatarsus) and leg (tibia) meet.

Hover Canopy used for brooder stoves to hold down the heat at bird level.

Husbandry Proper and timely care and management of livestock.

Incubate To maintain favorable conditions for the development and hatching of fertile eggs.

Keel bone Breastbone or sternum.

Keet A young guinea.

Lateral Relating to the sides.

Litter Soft, absorbent material used to cover floors of poultry houses.

Mandible The upper or lower bony portion of the beak.

Meleagris gallopavo The turkey, both wild and domestic.

Molt To shed old feathers, which are replaced by new ones.

Morbidity Illness.

NPIP National Poultry Improvement Plan, initiated by the USDA in 1935 to reduce transovarian diseases through hatchery sanitation and blood testing.

Numida meleagris Wild species of guinea fowl in Africa; domestic guineas bred from this species.

Offal Waste parts or entrails from butchered or processed birds and animals.

Oil sac or uropygial gland Large oil gland on the back at the base of the tail used by the bird to preen or condition its feathers.

Ova The yolks of eggs.

Ovary The female reproductive gland in which eggs are formed.

Oviduct Long glandular tube where egg formation takes place, leading from the ovary to the cloaca; it is made up of the funnel, magnum, isthmus, uterus, and vagina.

Oviposition The act of laying an egg.

Ovulation The release of the yolk from the ovary.

Ovum The female reproductive cell.

Parthenogenesis Initiation of cell division in an unfertilized egg without contribution from the male. Rarely ends in a hatched poult but it has occurred in turkeys. Offspring are predominantly male, with a small number of females.

Pendulous crop Crop that is impacted and enlarged and hangs down in an abnormal manner.

Pavo cristatus Indian peafowl.

Phasianus colchicus Ringnecked pheasant.

Pinioning Permanent removal of the outer wing joint to prevent fighting.

Pipping Young fowl breaking out of its shell.

Plumage The feathers making up the outer covering of birds.

Pores Thousands of minute openings in the shell of an egg through which gases are exchanged.

Posterior Toward the rear.

Poult A young turkey.

Poultry A term designating those species of birds used by humans for food or fiber that can be reproduced under their care. The term includes chickens, turkeys, ducks, geese, pheasants, and pigeons.

Preen gland Uropygial or oil gland found on the back near the tail that secretes oil for application on the feathers by the bird during preening.

Primary feathers The long, stiff flight feathers at the outer tip of the wing.

Protozoa Microscopic single-celled animals, such as those responsible for coccidiosis, histomoniasis (blackhead), trichomoniasis, and hexamitiasis.

Proventriculus True stomach of the bird located between the crop and the gizzard.

Range area An area of pasture or meadowland secured for livestock production.

Relative humidity The percentage of moisture saturation of the air; dependent on air temperature as well as the amount of moisture in the air.

Roost A perch on which birds rest or sleep.

Secondaries The large, stiff wing feathers adjacent to the body, visible when the wing is folded or extended.

Semen Fluid secreted by male reproductive organs; the vehicle for sperm transport.

Shank The scaly portion of the leg below the hock joint and between the thigh and the foot.

Shell The hard protective covering of an egg consisting primarily of calcium carbonate, secreted by the shell gland.

Shell gland That portion of the bird's reproductive tract (oviduct) where the shell and cuticle are deposited around the egg; also incorrectly referred to as the uterus.

Shell membranes The two soft fibrous membrane linings that surround the albumen; secreted in the isthmus. They normally separate at the large end of the egg to form an air cell.

Snood The fleshy appendage on the head of the turkey.

Sperm or spermatozoa The male reproductive cells capable of fertilizing the ova.

Squab A young pigeon from 1 to 30 days of age.

Squeaker A young pigeon from 30 days to 6 or 7 weeks of age.

Spur The stiff, horny structure on the legs of some birds; found on the inner side of the shank.

Strain Group of birds within a variety; fowl of any breed usually with the breeder's name that was reproduced by closed flock breeding for five or more generations.

Testes The male sex glands (plural)(testis = singular).

Tom The male turkey.

Trachea or windpipe That part of the respiratory system that conveys air from the larynx to the bronchi and to the lungs and air sacs.

Uterus Organ in female mammals in which the developing embryo is nourished; birds do not a uterus — see **shell gland.**

Vagina The section of the oviduct that holds the formed egg until it is laid; located between the shell gland and cloaca.

Variety A subdivision of breed usually distinguished by either color or color and pattern, also and/or comb type in chickens.

Vent or anus The external opening of the cloaca.

Ventral Of or relating to the lower part of the body, such as the breast or keel.

Vitelline membrane The membrane that surrounds the yolk.

Wet-bulb temperature Device to measure moisture or water vapor in the air.

Yolk Ovum, the yellow portion of the egg.

Index

Note: Page numbers in *italics* indicate illustrations; those in **boldface** indicate charts or tables.

Other Storey Titles You will Enjoy